Paraoxonases in Inflammation, Infection, and Toxicology

Advances in Experimental Medicine and Biology

Editorial Board:

Recent Volumes in this Series

Srinivasa T. Reddy
Editor

Paraoxonases in Inflammation, Infection, and Toxicology

 Humana Press

Contents

Contributors

Carlos Alonso-Villaverde Centre de Recerca Biomèdica, Hospital Universitari de Sant Joan, Institut d'Investigació Sanitària Pere Virgili, Universitat Rovira i Virgili, Reus, Spain, cavillaverde@grupsagessa.com

G.M. Anantharamaiah Departments of Medicine, Biochemistry and Molecular Genetics, University of Alabama at Birmingham, Birmingham, AL 35294, USA, ananth@uab.edu

Michael Aviram The Lipid Research Laboratory, Technion Faculty of Medicine, Rappaport Family Institute for Research in the Medical Science and Rambam Medical Center, Haifa 31096, Israel, aviram@tx.technion.ac.il

László Bajnok Department of Medicine, School of Medicine, University of Pécs, Pécs, Hungary, laszlo.bajnok@aok.pte.hu

Charles Bisgaier Esperion Therapeutics; University of Michigan, Ann Arbor, MI, USA, cbizzzzzzz@aol.com

Jordi Camps Centre de Recerca Biomèdica, Hospital Universitari de Sant Joan, Institut d'Investigació Sanitària Pere Virgili, Universitat Rovira i Virgili, Reus, Spain, jcamps@grupsagessa.cat

Douglas M. Cerasoli US Army Medical Research Institute of Chemical Defense, Aberdeen Proving Ground, MD, USA, douglas.cerasoli@us.army.mil

Manjula Chaddha Departments of Medicine, Biochemistry and Molecular Genetics, University of Alabama at Birmingham, Birmingham, AL 35294, USA, mchaddha@ab.edu

Harvey Checkoway Departments of Epidemiology; Departments of Environmental & Occupational Health Sciences, University of Washington, Seattle, WA, USA, checko@u.washington.edu

Toby B. Cole Departments of Medicine (Division of Medical Genetics) and Environmental and Occupational Health Sciences, tobycole@u.washington.edu

Lucio G. Costa Environmental and Occupational Health Sciences, University of Washington, Seattle, WA, USA, lgcosta@u.washington.edu

Geeta Datta Departments of Medicine, Biochemistry and Molecular Genetics, University of Alabama at Birmingham, Birmingham, AL 35294, USA, gdatta@uab.edu

Anneclaire J. De Roos Departments of Epidemiology, University of Washington; The Fred Hutchinson Cancer Research Center, Seattle, WA, USA, deroos@u.washington.edu

Sara P. Deakin Department of Internal Medicine, Faculty of Medicine, University of Geneva, Geneva, Switzerland, sara.deakin@hcuge.ch

Dragomir Draganov University of Michigan, Ann Arbor, MI, USA; WIL Research Laboratories, LLC, Ashland, OH, USA, ddraganov@wilresearch.com

Michal Efrat The Lipid Research Laboratory, Technion Faculty of Medicine, the Rappaport Family Institute for Research in the Medical Sciences and Rambam Medical Center, Haifa, Israel, michalef@gmail.com

M.L. Estin Department of Internal Medicine, University of Iowa, Iowa City, IA 52242 USA, miriam.estin@ucdenver.edu

Federico M. Farin Departments of Environmental & Occupational Health Sciences, University of Washington, Seattle, WA, USA, freddy@u.washington.edu

Richard A. Fenske Departments of Environmental & Occupational Health Sciences, University of Washington, Seattle, WA, USA, rfenske@u.washington.edu

Clement E. Furlong Departments of Medicine (Division of Medical Genetics) and Genome Sciences, University of Washington, Seattle, WA 98195, USA, clem@u.washington.edu

David W. Garber Departments of Medicine, Biochemistry and Molecular Genetics, University of Alabama at Birmingham, Birmingham, AL 35294, USA, dgarber@uab.edu

David R. Goodlett Department of Medicinal Chemistry, University of Washington, Seattle, WA, USA, goodlett@u.washington.edu

Himanshu Gupta Departments of Medicine, Biochemistry and Molecular Genetics, University of Alabama at Birmingham, Birmingham, AL 35294, USA, hgupta@uab.edu

Susan Hama David Geffen School of Medicine, University of California, Los Angeles, USA, shama@mednet.ucla.edu

Shaila P. Handattu Departments of Medicine, Biochemistry and Molecular Genetics, University of Alabama at Birmingham, Birmingham, AL 35294, USA, shandattu@uab.edu

Mariann Harangi Department of Internal Medicine, Medical and Health Science Centre, University of Debrecen, Debrecen, Hungary, mharangi@hotmail.com

Jonathan N. Hofmann Division of Cancer Epidemiology and Genetics, National Cancer Institute, Bethesda, MD, USA, hofmannjn@mail.nih.gov

Greg Hough David Geffen School of Medicine, University of California, Los Angeles, USA, ghough@mednet.ucla.edu

Hieronim Jakubowski Department of Microbiology & Molecular Genetics, UMDNJ-New Jersey Medical School, International Center for Public Health, Newark, NJ 07101, USA; Department of Biochemistry and Biotechnology, University of Life Sciences, Poznań, Poland, jakubows@umdnj.edu

Richard W. James Department of Internal Medicine, Faculty of Medicine, University of Geneva, Geneva, Switzerland, richard.james@hcuge.ch

Karen Jansen Environmental and Occupational Health Sciences, University of Washington, Seattle, WA, USA, kjansen@u.washington.edu

Gail P. Jarvik Departments of Medicine (Division of Medical Genetics) and Genome Sciences, University of Washington, Seattle, WA 98195, USA, pair@u.washington.edu

Jorge Joven Centre de Recerca Biomèdica, Hospital Universitari de Sant Joan, Institut d'Investigació Sanitària Pere Virgili, Universitat Rovira i Virgili, Reus, Spain, jjoven@grupsagessa.com

Matthew C. Keifer Departments of Environmental & Occupational Health Sciences, University of Washington, Seattle, WA, USA, mkeifer@u.washington.edu

J. Brian Kim David Geffen School of Medicine, University of California, Los Angeles, USA, jkim@mednet.ucla.edu

Jerry H. Kim Department of Anesthesiology, University of Washington, Seattle, WA, USA, jerkim@u.washington.edu

Peter Koncsos Department of Internal Medicine, Medical and Health Science Centre, University of Debrecen, Debrecen, Hungary, peterkoncsos@yahoo.com

Bert La Du University of Michigan, Ann Arbor, MI, USA

David E. Lenz US Army Medical Research Institute of Chemical Defense, Aberdeen Proving Ground, MD, USA, david.lenz@us.army.mil

Wan-Fen Li Environmental Health and Occupational Medicine, National Health Research Institutes, Zhunan Town, Taiwan, wanfenli@yahoo.com.tw

Aldons J. Lusis Division of Cardiology, David Geffen School of Medicine at UCLA, 10833 Le Conte Avenue, BH-307 CHS, Los Angeles, CA 90095-1679, USA, jlusis@mednet.ucla.edu

Nancy Ly David Geffen School of Medicine, University of California, Los Angeles, USA, nancyly@buffalo.edu

Michael J. MacCoss Department of Genome Sciences, University of Washington, Seattle, WA, USA, maccoss@u.washington.edu

Bharti Mackness Universitat Rovari i Virgili, Centre de Recerca Biomedica, Hospital Universitari de Sant Joan, 43201 Reus, Spain, bhartimackness@tiscali.co.uk

Mike Mackness Universitat Rovari i Virgili, Centre de Recerca Biomedica, Hospital Universitari de Sant Joan, 43201 Reus, Spain, mike.mackness@gmail.com

Judit Marsillach Centre de Recerca Biomèdica, Hospital Universitari de Sant Joan, Institut d'Investigació Sanitària Pere Virgili, Universitat Rovira i Virgili, Reus, Spain, jmarsillach@grupsagessa.com

Mohamad Navab David Geffen School of Medicine, University of California, Los Angeles, USA, mnavab@mednet.ucla.edu

Gaurav Nayyar Departments of Medicine, Biochemistry and Molecular Genetics, University of Alabama at Birmingham, Birmingham, AL 35294, USA, gnayyar@uab.edu

Jean Nemzek University of Michigan, Ann Arbor, MI, USA, jnemzek@umich.edu

Tamara C. Otto US Army Medical Research Institute of Chemical Defense, Aberdeen Proving Ground, MD, USA, tamara.c.otto@us.army.mil

Mayakonda N. Palgunachari Departments of Medicine, Biochemistry and Molecular Genetics, University of Alabama at Birmingham, Birmingham, AL 35294, USA, palgun@uab.edu

György Paragh Department of Internal Medicine, Medical and Health Science Centre, University of Debrecen, Debrecen, Hungary, paragh@internal.med.unideb.hu

Sarah Park Environmental and Occupational Health Sciences, University of Washington, Seattle, WA, USA; Departments of Medicine (Division of Medical Genetics) and Genome Sciences, Aberdeen Proving Ground, MD, USA, ssp29@u.washington.edu

Daniel Remick University of Michigan, Ann Arbor, MI, USA, remickd@bu.edu

Rebecca J. Richter Departments of Medicine (Division of Medical Genetics) and Genome Sciences, University of Washington, Seattle, WA 98195, USA; Aberdeen Proving Ground, MD, USA, rrichter@u.washington.edu

Anna Rull Centre de Recerca Biomèdica, Hospital Universitari de Sant Joan, Institut d'Investigació Sanitària Pere Virgili, Universitat Rovira i Virgili, Reus, Spain, arull@grupsagessa.com

Shila Safarpoor David Geffen School of Medicine, University of California, Los Angeles, USA, baharan50@yahoo.com

Alex Scherl Department of Medicinal Chemistry, University of Washington, Seattle, WA, USA, alexander.scherl@unige.ch

Ildikó Seres Department of Internal Medicine, Medical and Health Science Centre, University of Debrecen, Debrecen, Hungary, seres@internal.med.unideb.hu

Diana M. Shih Division of Cardiology, David Geffen School of Medicine at UCLA, 10833 Le Conte Avenue, BH-307 CHS, Los Angeles, CA 90095-1679, USA, dshih@mednet.ucla.edu

Theodore Standiford University of Michigan, Ann Arbor, MI, USA, tstandif@med.umich.edu

Richard C. Stevens Department of Medicine (Division of Medical Genetics) and Genome Sciences, University of Washington, Seattle, WA, USA; Aberdeen Proving Ground, MD, USA, rstevensboston@gmail.com

D.A. Stoltz Department of Internal Medicine, University of Iowa, Iowa City, IA 52242 USA, david-stoltz@uiowa.edu

Stephanie M. Suzuki Departments of Medicine (Division of Medical Genetics) and Genome Sciences, University of Washington, Seattle, Washington, USA, stephis@u.washington.edu

Ferenc Sztanek Department of Internal Medicine, Medical and Health Science Centre, University of Debrecen, Debrecen, Hungary, sztanekf@yahoo.com

Hagai Tavori The Lipid Research Laboratory, Technion Faculty of Medicine, Rappaport Family Institute for Research in the Medical Science and Rambam Medical Center, Haifa 31096, Israel; Laboratory of Natural Medicinal Compounds, MIGAL – Galilee Technology Center, P.O. Box 831, Kiryat Shmona 11016, and Tel Hai College, Israel, hagait@migal.org.il

John Teiber University of Michigan, Ann Arbor, MI, USA; The University of Texas Southwestern Medical Center, Dallas, TX, USA, john.teiber@utsouthwestern.edu

Duc Tien David Geffen School of Medicine, University of California, Los Angeles, USA, duc.anh.tien@gmail.com

Ladan Vakili David Geffen School of Medicine, University of California, Los Angeles, USA, lvakili@mednet.ucla.edu

Ghazal Vakili David Geffen School of Medicine, University of California, Los Angeles, USA, ghvakilimd@yahoo.com

Gerald van Belle Departments of Environmental & Occupational Health Sciences; Departments of Biostatistics, University of Washington, Seattle, WA, USA, vanbelle@u.washington.edu

Jacob Vaya Laboratory of Natural Medicinal Compounds, MIGAL - Galilee Technology Center, P.O. Box 831, Kiryat Shmona 11016, and Tel Hai College, Israel, vaya@migal.org.il

Catherine Watson Esperion Therapeutics, Ann Arbor, MI, USA, cwatson@umich.edu

C. Roger White Departments of Medicine, Biochemistry and Molecular Genetics, University of Alabama at Birmingham, Birmingham, AL 35294, USA, crwhite@uab.edu

Yu-Rong Xia Division of Cardiology, David Geffen School of Medicine at UCLA, 10833 Le Conte Avenue, BH-307 CHS, Los Angeles, CA 90095-1679, USA, yurangxia@mednet.ucla.edu

Janet M. Yu Division of Cardiology, David Geffen School of Medicine at UCLA, 10833 Le Conte Avenue, BH-307 CHS, Los Angeles, CA 90095-1679, USA, janetyu@ucla.edu

J. Zabner Department of Internal Medicine, University of Iowa, Iowa City, IA 52242 USA, joseph-zabner@uiowa.edu

The 3rd International Conference on Paraoxonases (ICP) was a great success. More than 100 basic and clinical researchers working on the paraoxonase (PON) family of proteins (PON1, PON2, and PON3) gathered at the University of California, Los Angeles, USA, located in beautiful, sunny Southern California, from September 7th to 10th, 2008, and presented their findings in the area of paraoxonases. The conference included expert lectures from 21 invited speakers, 20 short oral presentations, and around 60 poster presentations. The 3rd ICP provided an excellent environment for cross-fertilization of ideas among all the researchers in the field.

Over the last 5 years PON research has grown exponentially. It is now well established that PON proteins play important roles in inflammation, infection, and toxicology. The lectures and the short presentations at the 3rd ICP covered areas that included research on the genetics, biochemistry, structural biology, and regulation of PON genes. The scientific committee of the 3rd ICP unanimously agreed to put several important reports from the conference into a book that will aid the research community to keep pace with this ever-growing area of research that has a great deal of promise in the fight against inflammatory diseases, infectious diseases, and toxins.

ApoE Mimetic Peptide Reduces Plasma Lipid Hydroperoxide Content with a Concomitant Increase in HDL Paraoxonase Activity

Geeta Datta, Manjula Chaddha, Shaila P. Handattu, Mayakonda N. Palgunachari, Gaurav Nayyar, David W. Garber, Himanshu Gupta, C. Roger White, and G.M. Anantharamaiah

Abstract ApoE mimetic peptide possesses the putative receptor binding domain 141–150 (LRKLRKRLLR) of apoE covalently linked to the class A amphipathic helical peptide 18A. It dramatically reduces plasma cholesterol in dyslipidemic mouse and rabbit models. Recycling of apoE mimetic peptide increases the duration of preβ-HDL formation leading to extended anti-inflammatory and atheroprotective properties.

Keywords ApoE mimetic · Preβ-HDL · Lipid hydroperoxide

The apoE mimetic peptide is composed of the putative receptor binding domain of apoE (residues 141–150, LRKLRKRLLR) covalently linked to the class A amphipathic helical peptide 18A. The resulting peptide, with both the amino and carboxyl termini blocked, is called Ac-hE18A-NH$_2$ (Datta et al. 2000). It enhances the uptake of apoB-containing lipoproteins in HepG2 cells via heparan sulfate proteoglycans (HSPG) pathway. When administered to dyslipidemic mouse models, it dramatically enhances the hepatic uptake of atherogenic lipoproteins and reduces plasma cholesterol (Garber et al. 2003). We have shown previously that in the Watanabe Heritable Hyperlipidemic (WHHL) rabbit, a single administration of Ac-hE18A-NH$_2$ decreased plasma lipid hydroperoxide levels from 25 to 12 μM in 2 hours. Concomitantly, paraoxanase (PON) activity (Gupta et al. 2005) increased from 150 to 800 U.

In NZW rabbits on a 1% cholesterol diet, saline administered rabbits exhibited an increase in plasma cholesterol from 625 to 1300 mg/dL in 6 days (Fig. 1). However, this increase was not seen in rabbits (on a 1% cholesterol diet) administered a single dose (7.5 mg/kg) of peptide for the same time period. Interestingly, this effect persisted for 7 days even though the half-life of the peptide is very short (Garber et al. 2003). A second administration of the peptide after 7 days maintained

G. Datta (✉)
Departments of Medicine, Biochemistry and Molecular Genetics, University of Alabama at Birmingham, Birmingham, AL, 35294, USA
e-mail: gdatta@uab.edu

S.T. Reddy (ed.), *Paraoxonases in Inflammation, Infection, and Toxicology*, Advances in Experimental Medicine and Biology 660, DOI 10.1007/978-1-60761-350-3_1,
© Humana Press, a part of Springer Science+Business Media, LLC 2010

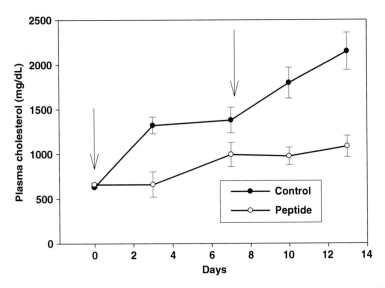

Fig. 1 The plasma cholesterol-lowering effect of Ac-hE18A-NH₂ is sustained over a period of 7 days in NZW rabbits. NZW rabbits ($n = 4$) on a 1% cholesterol diet were administered saline (control group; *filled circles*) or Ac-hE18A-NH₂ (7.5 mg/kg i.v.; experimental group; *open circles*). The *arrows* indicate the day on which the peptide was administered. Plasma cholesterol in control rabbits kept rising with time while the peptide-administered rabbits had significantly lower plasma cholesterol levels even 7 days after the peptide administration ($p < 0.05$)

plasma cholesterol levels at 1000 mg/dL while the saline controls had increased to 2100 mg/dL.

To better understand such a dramatic long-lasting cholesterol-reducing property, we hypothesized that Ac-he18A-NH₂ is internalized and re-released at later time points. Immunohistochemical analysis of HepG2 cells incubated with Ac-hE18A-NH₂ demonstrated that the peptide is internalized by HepG2 cells (Fig. 2). The cells were viable under these conditions as measured by trypan blue exclusion. Pulse chase studies with [¹²⁵I]-Ac-hE18A-NH₂ showed that the peptide is taken up and is secreted at later time points (Fig. 3), suggestive of recycling of the peptide, similar to that observed with apoE. This recycling phenomenon explains the long-lasting cholesterol-reducing effect observed in NZW rabbits (Fig. 1).

In HepG2 cells, Ac-hE18A-NH₂ promotes the secretion of apoA-I as a lipid-poor preβ-HDL particle (Fig. 4). This dual-domain peptide can displace apoA-I from HDL to form a lipid-poor preβ-like HDL particle. To determine if secretion of preβ-HDL particles in peptide-treated samples is affected after removal of the peptide, the peptide-containing conditioned media was replaced with fresh medium after an overnight incubation. Media from cells that were treated with peptide showed preβ-HDL particles in spite of the fact that there was no peptide added. This suggested that the peptide that had been internalized was secreted, supporting the recycling of the peptide. Recycling of the apoE mimetic peptide increases the duration of

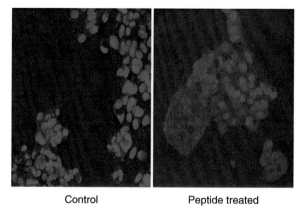

Control   Peptide treated

Fig. 2 Ac-hE18A-NH$_2$ enters HepG2 cells. HepG2 cells were incubated with MEM containing Ac-hE18A-NH$_2$ (20 μg/ml) for 1 h. Control cells were incubated with MEM without the peptide. Cells were immunostained with antibody to Ac-hE18A-NH$_2$ and viewed under a confocal microscope. The nuclei were stained with Hoechst (*blue*) and the peptide was visualized by rhodamine-labeled secondary antibody (*red*). No non-specific labeling was observed

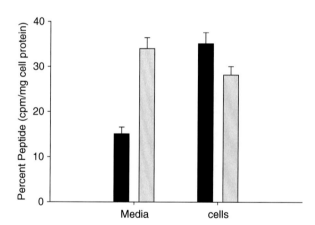

Fig. 3 The peptide, Ac-hE18A-NH$_2$, recycles back to the surface of HepG2 cells and is secreted into the medium. The *dark bars* represent the percent of peptide 5 min after incubation and the grey bars represent the percent of peptide 60 min after incubation. The experiment was repeated three times with $n = 6$ each time

preβ-HDL formation leading to extended anti-inflammatory and atheroprotective properties (Fig. 4).

It is thought that HDL is a ligand for carrying PON from hepatocytes. We are in the process of understanding whether the peptide increases HDL and/or forms HDL-like particles with concomitant increase in PON levels.

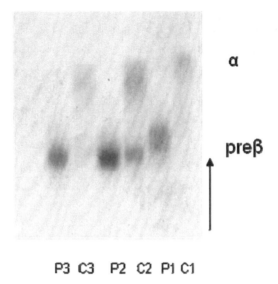

P3 C3 P2 C2 P1 C1

Fig. 4 The effect of Ac-hE18A-NH$_2$ on the formation of preβ-HDL is seen even after removal of the peptide from the incubation medium. HepG2 cells (at 90% confluency) were incubated at 37°C with or without Ac-hE18ANH$_2$ (20 μg/ml). After an overnight incubation (lanes C1 and P1), the conditioned medium was removed and fresh medium *without the peptide* was added both to control as well as to cells that had previously been treated with the peptide. The medium was then removed and subjected to agarose gel electrophoresis. Proteins were immunoblotted for apoA-I. Lanes P1 (peptide-treated) and C1 (control) show the formation of preβ-HDL in the presence of the peptide. Lanes P2 and P3 (peptide-treated) show the preβ-HDL subspecies formed in the medium of cells that had been previously exposed to Ac-hE18A-NH$_2$ for 18 h followed by 24 and 48 h incubation respectively. In sharp contrast, lanes C2 and C3 (control) show that the control cells contain predominantly α-HDL subspecies 24 and 48 h after changing the medium

References

Datta, G., M. Chaddha, D.W.Garber, B.H.Chung, E.M.Tytler, N. Dashti, W.A.Bradley, S.H.Gianturco, and G.M.Anantharamaiah. 2000. The receptor binding domain of apolipoprotein E, linked to a model class A amphipathic helix, enhances internalization and degradation of LDL by fibroblasts. *Biochemistry* **39**: 213–220

Garber, D.W., S.Handattu, I. Aslan, G.Datta, M. Chaddha, and G.M.Anantharamaiah. 2003. Effect of an arginine-rich helical peptide on plasma cholesterol in dyslipidemic mice. *Atherosclerosis* **168**: 229–237

Gupta, H., C.R. White, S. Handattu, D.W. Garber, G. Datta, M. Chaddha, L. Dai, S.H. Gianturco, W.A. Bradley, and G.M. Anantharamaiah. 2005. Apolipoprotein E mimetic peptide dramatically lowers plasma cholesterol and restores endothelial function in Watanabe Heritable Hyperlipidemic rabbits. *Circulation* **111**: 3112–3118

Interrelationships Between Paraoxonase-1 and Monocyte Chemoattractant Protein-1 in the Regulation of Hepatic Inflammation

Jordi Camps, Judit Marsillach, Anna Rull, Carlos Alonso-Villaverde, and Jorge Joven

Abstract Oxidative stress and inflammation play a central role in the onset and development of liver diseases irrespective of the agent causing the hepatic impairment. The monocyte chemoattractant protein-1 is intimately involved in the inflammatory reaction and is directly correlated with the degree of hepatic inflammation in patients with chronic liver disease. Recent studies showed that hepatic paraoxonase-1 may counteract the production of the monocyte chemoattractant protein-1, thus playing an anti-inflammatory role. The current review summarises experiments suggesting how paraoxonase-1 activity and expression are altered in liver diseases, and their relationships with the monocyte chemoattractant protein-1 and inflammation.

Keywords Inflammation · Liver impairment · Monocyte · Chemoattractant protein-1 · Oxidative stress · Paraoxonase-1

Abbreviations

ABCA1	ATP-binding cassette transporter 1
ALD	alcoholic liver disease
CAM	cell adhesion molecules
HCV	hepatitis C virus
HDL	high-density lipoproteins
LDL	low-density lipoproteins
MCP-1	monocyte chemoattractant protein-1
NAFLD	non-alcoholic fatty liver disease
NASH	non-alcoholic steatohepatitis
PON1	paraoxonase-1
PPAR	peroxisome proliferator-activated receptor
ROS	reactive oxygen species
TBBL	5-thiobutyl butyrolactone

J. Camps (✉)
Centre de Recerca Biomèdica, Hospital Universitari de Sant Joan, C. Sant Joan s/n, 43201-Reus, Catalunya, Spain
e-mail: jcamps@grupsagessa.cat

S.T. Reddy (ed.), *Paraoxonases in Inflammation, Infection, and Toxicology*, Advances in Experimental Medicine and Biology 660, DOI 10.1007/978-1-60761-350-3_2,
© Humana Press, a part of Springer Science+Business Media, LLC 2010

1 Oxidative Stress in Chronic Liver Impairment and Its Relationship to Inflammation and Fibrosis

Increased oxidative stress and inflammation play a fundamental role in the onset and development of liver diseases. In this section we are going to show that the molecular mechanisms underlying the morphological and functional alterations observed in liver diseases present many points in common, independently of the agent causing the hepatic impairment. The most important causes of chronic liver disease in Western societies are alcohol abuse, obesity, and hepatitis C virus infection.

Alcoholic liver disease (ALD) encompasses a broad spectrum of hepatic alterations ranging from steatosis and minimal injury to advanced fibrosis and cirrhosis (Day, 2006). The involvement of oxidative injury in ethanol toxicity has emerged from reports showing that alcohol-fed animals and patients with ALD present with high content of lipid peroxidation products in their livers and in their circulation (Albano, 2006). Experimental studies demonstrated that preventing lipid peroxidation in ethanol-fed animals, by modulating the dietary content of polyunsaturated fatty acids, antioxidants, or inhibitors of free radical generation, reduces focal necrosis and inflammation (Nanji, 2004). Oxidative stress associated with ethanol intake comes mainly from reactive oxygen species (ROS) generated by the mitochondrial respiratory chain and cytochrome P4502E1 from hepatocytes, and the NADPH oxidase from Kupffer cells and recruited macrophages (Albano, 2008). The impairment of mitochondrial lipid oxidation is one of the mechanisms responsible for hepatic fat accumulation (Pessayre and Fromenty, 2005). Deletions in mitochondrial DNA show a high prevalence in patients with steatosis, a lesion that is ascribed to alterations in mitochondrial β-oxidation of fatty acids (Fromenty et al. 1995). Pan et al. (2004) reported that lipid peroxidation reduces hepatic lipoprotein secretion by enhancing the degradation of newly synthesised apolipoproteins and this effect, together with alterations in lipoprotein glycosylation in the Golgi apparatus (Albano, 2006), might contribute to microvesicular steatosis. Further evidence suggests that alcohol-induced oxidative stress interferes with the regulation of lipid synthesis by the peroxisome proliferator-activated receptor (PPAR)-α and the sterol regulatory element binding protein 1 (Crabb and Liangpunsakul, 2006).

The possible role of oxidative stress in promoting an inflammatory reaction in ALD has emerged from the observation that lipid peroxidation end-products are able to develop an immune response (Mottaran et al. 2002; Thiele et al. 2004), and that liver-associated lymphocytes isolated from ethanol-fed rats have an increased capacity to secrete pro-inflammatory cytokines (Batey et al. 2002). Ethanol-induced lipid peroxidation also increases the production of the pro-fibrogenetic tissue growth factor β-1 by Kupffer cells (Tsukamoto and Lu, 2001) and the expression of the collagen $\alpha2(1)$ gene by activated stellate cells (Nieto, 2007).

Non-alcoholic fatty liver disease (NAFLD) and non-alcoholic steatohepatitis (NASH) are hepatic lesions that appear frequently in obese and diabetic individuals despite the fact that they may not have a history of alcohol abuse (Solís Herruzo et al. 2006). These lesions resemble those of ALD, and are characterised by steatosis, hepatocyte hydropic degeneration, and inflammatory infiltrates. In addition, alterations in mictochondrial shape and function, and fibrosis in varying degrees, are usually found (Kleiner et al. 2005). NAFLD is a common lesion in Western populations, and will become more so in the future, as it is associated with insulin resistance, metabolic syndrome, diabetes, and obesity. It is estimated that 17–33% of the general population in Europe and the USA have NAFLD (Younossi et al. 2002). Oxidative stress plays a pivotal role in the evolution from "benign" steatosis to the more severe NASH. The "two hits" theory postulates that NASH would result from two challenges. The first one would be represented by fatty liver; the second by oxidative stress (Chittury and Farrell, 2001). While the source of oxidative stress in NASH is probably compound and derived from the metabolic disturbances associated with obesity, diabetes, etc., mitochondrial dysfunction seems to play an important role. Several studies have shown that mitochondria in patients with NASH are abnormal from both the morphological and the functional points of view and, as in ALD, alterations in the fatty acid β-oxidation promote an increased free radical production and lipid peroxidation (Solís Herruzo et al. 2006). The consequences of oxidative stress in NASH would be similar to those of ALD, with altered lipoprotein synthesis and secretion, an inflammatory reaction, and fibrosis.

Hepatitis C virus (HCV) is a major cause of viral hepatitis. In the USA alone about four million people are infected, and 35,000 new HCV cases are estimated to occur every year (Choi and Ou, 2006). Infection by this virus frequently does not resolve, and about 80% of the infected individuals become chronic carriers who may then progress to the most severe forms of liver impairment, as cirrhosis or hepatocellular carcinoma. HCV infection is characterised by increased oxidative stress (Choi et al. 2004). Lipid peroxidation products, aldehydes as 4-hydroxynonenal, and 8-hydroxyguanosine (a marker of oxidative DNA damage) are elevated in these patients (Mahmood et al. 2004). The increased oxidative stress may be explained by chronic inflammation and the generation of free radicals by Kupffer cells and recruited macrophages (Forman et al. 2003). NS3 protein of HCV has been found to activate Nox 2 protein from macrophages, leading to increased generation of ROS that can exert oxidative stress on the nearby cells (Thoren et al. 2004). Furthermore, studies have indicated that HCV can directly induce oxidative stress in hepatocytes. HCV core gene expression has been associated with increased ROS, decreased reduced glutathione content, and increased thioredoxin in parenchymal cells (Abdallah et al. 2005). Recent studies showed that HCV core proteins bind to the outer mitochondrial membrane resulting in mitochondrial dysfunction by Ca^{2+} accumulation. These alterations would inhibit electron transport and promote ROS production (Choi and Ou, 2006). Another HCV protein, NS5A, has also been reported to increase free radical production by Huh7 cells (Tardif et al. 2005). As in ALD and NASH, increased oxidative stress would produce a multifactorial reaction involving the synthesis of pro-inflammatory and pro-fibrogenetic cytokines and chemokines.

Therefore, it seems evident that chronic liver diseases share common biochemical alterations irrespectively of their aetiology. They are all accompanied by an

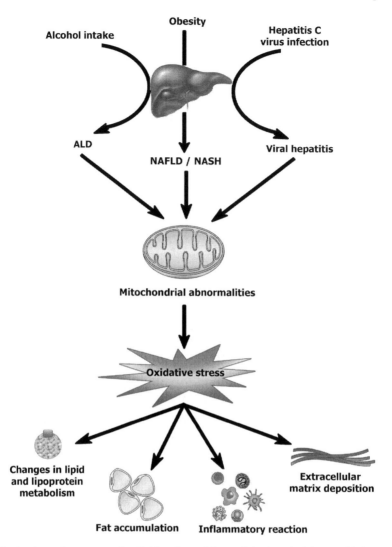

Fig. 1 A schematic representation linking the aetiology of the hepatic injury, oxidative stress, and biochemical and histological alterations in chronic liver diseases. Alcoholic liver disease (ALD), non-alcoholic fatty liver disease/steatohepatitis (NAFLD/NASH) and hepatitis C virus infection promote an increase in free radical production in the hepatocyte mitochondria, inducing changes in the intracellular fat accumulation, lipoprotein metabolism, an inflammatory reaction, and extracellular matrix synthesis

increased oxidative stress secondary to mitochondrial abnormalities, promoting changes in lipid and lipoprotein metabolism, fat accumulation, an exacerbation of the inflammatory reaction due to increased cytokine synthesis, and extracellular matrix deposition (Fig. 1).

2 Monocyte Chemoattractant Protein-1 (MCP-1) and the Regulation of Hepatic Inflammation

Extensive experimental data show a link between lipid peroxidation and inflammation (Navab et al. 2004). High-density lipoprotein (HDL) particles possess antioxidant (Navab et al. 2000a, b; Negre-Salvayre et al. 2006) and anti-inflammatory properties, including the suppression of cytokine-induced endothelial cell adhesion molecules (CAM) (Cockerill et al. 1995; Clay et al. 2001; Calabresi et al. 2002). Recent studies in healthy volunteers showed a close relationship between HDL levels and the inflammatory response to endotoxin challenge; the incidence and severity of clinical symptoms and the plasma concentrations of tumour necrosis factor, interleukins 1, 6 and 8, and MCP-1 being higher in subjects with low HDL than in those with normal HDL levels (Birjmohun et al. 2007).

MCP-1 is intimately involved in the inflammatory reaction. This chemokine regulates the migration of monocytes into tissues and their subsequent differentiation into macrophages (Simpson et al. 2003). An end-product of lipid peroxidation (4-hydroxy-2-nonenal) and at least two oxidised phospholipids present in oxidised LDL [1-palmitoyl-2-(5-oxovaleroyl)-sn-glycero-3-phosphorylcholine and 1-palmitoyl-2-glutaroyl-sn-glycero-3-phosphorylcholine] have been shown to stimulate the production of MCP-1 in vitro (Parola et al. 1999; Navab et al. 2001; Zamara et al. 2004). Moreover, HDL has been shown to attenuate the stimulation of monocyte migration induced by oxidised LDL (Navab et al. 1991) in co-cultures of human monocytes and endothelial cells, providing indirect evidence that HDL suppresses MCP-1 production.

Liver diseases are inflammatory processes associated with increased plasma concentrations of cytokines and chemokines. An enhanced expression of MCP-1 was observed in liver parenchymal cells at sites of inflammation in patients with alcoholic hepatitis (Afford et al. 1998; Devalaraja et al. 1999), and MCP-1 concentrations were also found increased in the peripheral and the hepatic veins of patients with this disease (Fisher et al. 1999). These results suggested that MCP-1 may play an important role in the stimulation of the inflammatory infiltrate, and might also have immunomodulatory effects, including enhanced expression of adhesion molecules in monocytes and promotion of a pro-inflammatory cytokine synthesis, thus amplifying the inflammatory cascade (Jiang et al. 1992). Plasma MCP-1 concentrations are directly correlated with the degree of hepatic inflammation in patients with chronic liver disease (Fig. 2), and its measurement has been proposed as a non-invasive index to evaluate this derangement (Marsillach et al. 2005). In several experimental models, a significant relationship has been demonstrated between MCP-1 and hepatic inflammation. Apolipoprotein E-deficient mice given a high-fat, high-cholesterol diet develop hepatic inflammation with a concomitant increase in the hepatic expression of *MCP-1* gene (Tous et al. 2006). In obese mice, leptin administration induced MCP-1 production by stellate cells and hepatic inflammation (Aleffi et al. 2005). In addition, anti-*MCP-1* gene therapy has been shown to prevent the development of hepatic fibrosis in dimethylnitrosamine-administered

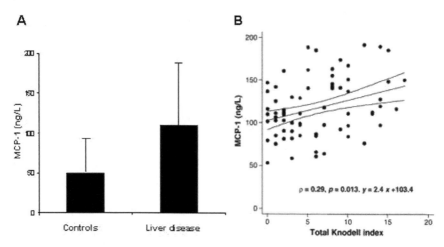

Fig. 2 (A) Plasma MCP-1 concentration in patients with chronic liver impairment and in control subjects. (B) Relationship between the Knodell index and plasma MCP-1 concentrations in patients with chronic liver impairment. The panel in (B) is reproduced from Marsillach et al. (2005) with permission. Copyright: Elsevier, 2005

rats (Tsuruta et al. 2004), while inhibition of CCR2 (a MCP-1 receptor) with propagermanium ameliorates insulin resistance and hepatic steatosis in diabetic mice (Tamura et al. 2008).

3 The Antioxidant and Hepatoprotective Function of Paraoxonase-1 (PON1) in Chronic Liver Impairment

In chronic liver diseases, oxidative stress influences the pathophysiological changes leading to liver cirrhosis and to hepatocellular carcinoma. Since PON1 exerts a protective effect against oxidative stress, it is logical to find an association between this enzyme and liver impairment. Ferré et al. (2001) observed, in rats with carbon tetrachloride-induced fibrosis, that an inhibition of liver microsomal PON1 activity was an early biochemical change related to increased lipid peroxidation and liver damage. They investigated the relationships between hepatic microsomal PON1 activity, lipid peroxidation and the progress of the disease in this experimental model. Moreover, they monitored the modulation of these processes by a dietary supplementation of zinc, a metal that possesses antioxidant and anti-fibrogenetic properties. They found that PON1 activity decreased while lipid peroxidation increased in carbon tetrachloride-administered rats, while the addition of zinc was associated with enhanced PON1 activity and a normalisation of lipid peroxidation. This study suggested that PON1 activity may be involved in the defence against free radical production in liver organelles.

Studies between the 1970s and the 1990s observed for the first time a significant decrease in serum PON1 activity in small groups of patients with liver cirrhosis

(Burlina and Galzigna, 1974; Burlina et al. 1977; Kawai et al. 1990). This results were confirmed by Ferré et al. (2002, 2005) in a wider series of patients with various degrees of chronic liver damage. These latter studies noted a significant decrease of serum PON1 activity in patients with chronic hepatitis, and an even greater decrease in cirrhotic patients, compared to a control group. Similar results were obtained by Kilic et al. (2005) in patients with chronic hepatitis.

A genetic association between the *PON1R* allele from the *PON1$_{192}$* polymorphism and chronic hepatitis C virus infection has been observed (Ferré et al. 2005). A possible explanation for this finding could be found in the fact that the *PON1Q* allele seems to be more efficient than the *PON1R* allele in hydrolysing lipid peroxides (Aviram et al. 2000). In support of this hypothesis, the total plasma peroxide concentrations were significantly increased in patients with chronic hepatitis, and the concentrations were highest in subjects who carried the RR genotype.

The effects of alcohol intake on serum PON1 levels depend on the degree of liver dysfunction. Studies have suggested that PON1 genetic polymorphisms are not a major factor in the modulation of serum PON1 levels consequent to alcohol consumption (Costa et al. 2005). To date, there has been only one study conducted in chronic alcohol abusers. The subjects were classified into several sub-groups according to their degree of liver disease. The results demonstrated that serum PON1 activity was decreased in alcoholic patients, and that the magnitude of the alteration was related to the degree of liver damage (Marsillach et al. 2007). These findings differ from those described in normal volunteers reporting moderate alcohol consumption, in whom serum PON1 activity and HDL cholesterol were found to be slightly increased (Rao et al. 2003). Patients with chronic alcoholic abuse and null or minimal hepatic change were observed to have an inverse relationship between PON1 activity and plasma malondialdehyde concentration.

Several mechanisms may be proposed to explain the decrease of serum PON1 activity in chronic liver diseases. First, these patients present with an increased free radical production, and it has been previously reported that PON1 is inactivated after hydrolysing lipid peroxides (Aviram et al. 1999). Second, alterations in HDL structure and composition can affect PON1 activity (Deakin et al. 2002; James and Deakin, 2004). Third, PON1 activity may be decreased secondarily to an altered PON1 synthesis by the liver.

Serum PON1 measurement has been proposed as an useful test for the evaluation of the degree of liver impairment. Since standard biochemical tests for liver dysfunction are insufficiently sensitive for a reliable indication of presence or absence of liver disease, histological examination of liver biopsy material has become the diagnostic tool of choice. However, finding new, non-invasive, reliable tests is an ongoing field of research that has considerable clinical value. Some studies suggest that the measurement of serum PON1 activity may add valuable information in the assessment of the extent of liver damage. Serum PON1 activity has a high diagnostic accuracy when distinguishing patients with liver disease from control subjects and, when added to a standard battery of liver function tests, increases the overall sensitivity without impairing the specificity (Ferré et al. 2002; Marsillach et al. 2007; Camps et al. 2007). The main drawback is that the use of toxic and unstable substrates to measure PON1 activity hampers the implementation of this assay in routine practice. However, the recent development of new assays for serum PON1 measurement by using non-toxic substrates, such as the lactonase assay employing 5-thiobutyl butyrolactone (TBBL), makes this proposal closer to a practical development (Marsillach et al. 2008).

The value of measuring serum PON1 activity has been studied in relation to outcomes of liver transplantation in patients with severe liver disease (Xu et al. 2005). The serum PON1 activity was low, but tended to increase, in liver-transplanted patients when the hepatic arteries had become blocked. Since PON1 activity is closely related to the recovery of liver function, its measurement could provide more accurate information on the success, or otherwise, of the liver transplant.

Further studies showed that PON1 concentrations in serum and liver homogenates were not decreased, but were significantly increased in patients with liver disease, and that this increase was related to an inhibition of apoptosis in liver parenchymal cells (Ferré et al. 2006). Recent evidence indicates that HDL inhibits cell apoptosis and that oxidised HDL loses this capability (Sugano et al. 2000; Nofer et al. 2002). It is likely that PON1, which is known to protect HDL from oxidation, would encourage the anti-apoptotic potential of HDL. Ferré et al. (2006) observed that enhanced PON1 protein expression was associated with an increased serum soluble Fas concentration (which is a marker of anti-apoptosis), together with decreased Fas-positive cell clusters and parenchymal cell DNA fragmentation (both markers of apoptosis) in the liver biopsies of patients with liver disease. These patients had also a higher serum PON1 concentration and protein expression in the liver. These data suggest that PON1 influences the anti-apoptotic capability of the HDL molecule, perhaps because of its capacity to protect the lipoprotein against oxidation.

4 Hepatic PON1 Synthesis and Its Relationships with MCP-1 and Inflammation in Chronic Liver Disease

What are the mechanisms underlying the alterations in serum PON1 activity and concentration and PON1 protein expression in liver impairment? Further studies, in rats with CCl_4-induced cirrhosis, concluded that increased PON1 concentrations in serum and liver homogenates were accompanied by a concomitant inhibition of *PON1* gene expression. These results were apparently contradictory, but we have to keep in mind that the intracellular accumulation of any protein is a result of a balance between its synthesis, its degradation, and its secretion to the medium. Several nuclear transcription factors have been reported to regulate PON1 synthesis: Sp1 and Ap-1, and the peroxisome proliferator-activated receptor (PPAR) family (Deakin et al. 2003; Fürnsinn et al. 2007). Rats with CCl_4-induced cirrhosis had a significant decrease in *PPARδ* gene expression and in one of the components of the Ap-1 family (Fra-2), and *PPARδ* gene expression was significantly correlated with *PON1* gene expression.

Hepatic *PON1* gene expression decrease in liver disease seems, then, associated with a decrease in PPARδ and Ap-1 activity. However, why is PON1 protein expression increased? The same studies demonstrated that rats with CCl$_4$-induced liver impairment had a decreased cathepsin B activity. This enzyme is a lysosomal cysteine protease and its levels are a reliable index of the overall hepatic proteolysis (Kharbanda et al. 1995). These results suggested that hepatic PON1 protein degradation is inhibited in this experimental model. In addition, the decrease in *PPARδ* gene expression suggests that PON1 secretion to HDL is also hampered. PPARδ increases apolipoprotein A-I and HDL synthesis by mechanisms involving activation of the ATP-binding cassette transporter 1 (*ABCA1*) gene (Fürnsinn et al. 2007). The decrease in *PPARδ* gene expression observed in this study would accord with an inhibition of the hepatic HDL synthesis and a decreased PON1 secretion into the circulation. A precise explanation for *PPARδ* gene expression decrease in experimental liver disease is difficult. This transcription factor inhibits apoptosis in cultured cardiomyoblasts (Pesant et al. 2006) and hepatocyte apoptosis previously been described as being stimulated in experimental liver disease (Cabré et al. 1999). Probably, PPARδ levels decrease as a part of a pro-apoptotic mechanism tending to eliminate damaged hepatocytes during the process of liver damage. Summarising these experiments, it can be suggested that despite *PON1* gene expression being decreased in experimental liver diasese, the intrahepatic PON1 protein expression is increased, secondary to an inhibited degradation and secretion.

Is enhanced hepatic PON1 concentration just an epiphenomenon, or does it play any protective role against inflammation? Mackness et al (2004) demonstrated that PON1 inhibits MCP-1 production in cultured endothelial cells incubated with oxidised LDL. These authors found that recombinant PON1 abolished MCP-1 production in endothelial cells, while avian HDL (which, unlike human HDL, does not have PON1 bound with it) was unable to elicit this reaction. PON1 inhibition of MCP-1 production appeared to be due to its capacity to inhibit LDL oxidation.

Immunochemical studies showed that the hepatic expressions of PON1 and MCP-1 were significantly increased in CCl$_4$-administered rats (Fig. 3a, b), and that the distribution was similar for both proteins (Marsillach et al. 2009). Hepatocytes were strongly stained for PON1 and MCP-1 when located in close proximity to inflammatory infiltrates and fibrosis septa. The amounts of positively-stained hepatocytes were lower in the less-affected areas of the liver. Figure 3c shows the concentrations of PON1 and MCP-1 in liver homogenates of CCl$_4$-administered rats. Hepatic PON1 concentration showed a significant increase after 6 and 8 weeks of the hepatotoxin administration, and a further trend towards a decrease at week 12. When PON1 concentration was high, hepatic MCP-1 concentration remained low. This parameter only increased after 12 weeks of CCl$_4$ administration, when PON1 begun to decrease. These results suggest that PON1 in liver parenchymal cells may act as a barrier against MCP-1 production, and only when this barrier is overcome does this chemokine increase. These experiments add an in vivo support to the findings by Mackness et al (2004) in endothelial cells, and suggest that similar relationships between PON1 and MCP-1 may develop in the liver.

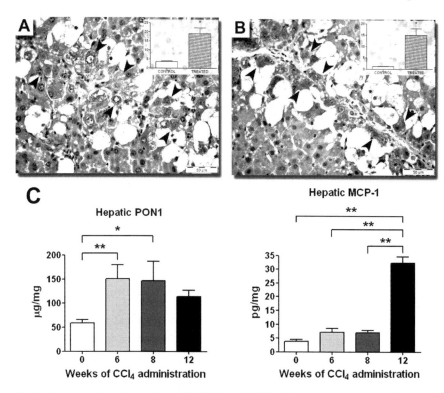

Fig. 3 Representative micrographs of PON1 (A) and MCP-1 (B) protein expression in liver tissue of rats treated with CCl$_4$ for 12 weeks. Original magnification: ×100. The inserts represent the mean percentages of positively-stained areas with respect to the total, in control and treated animals. *$p < 0.001$. (C): Changes in hepatic PON1 and MCP-1 concentration in CCl$_4$-administered rats up to 12 weeks

5 Conclusion

The studies shown in the present review suggest an active protective role for PON1 against oxidative stress-induced inflammation and fibrosis in the liver. Further investigations with animals treated with recombinant PON1, artificial PON1-containing lipid constructs or with PPARδ agonists would be valuable in investigating measures to counteract the inflammatory and fibrogenetic processes in chronic liver impairment and, perhaps, to provide new effective tools to treat this disease.

Acknowledgments Some studies described in this article have been funded by grants from the *Instituto de Salud Carlos III* (FIS 02/0430, 04/1752, 05/1607, RCMN C03/08 and RD06), *Ministerio de Sanidad*, Madrid, Spain. J. M. is the recipient of a post-graduate fellowship from the *Generalitat de Catalunya* (FI 05/00068).

References

Abdallah MY, Ahmad IM, Spitz DR et al. (2005) Hepatitis C virus-core and non structural proteins lead to different effects on cellular antioxidant defenses. J Med Virol 76:489–497

Afford SC, Fischer NC, Neil DAH et al. (1998) Distinct patterns of chemokine expression are associated with leukocyte recruitment in alcoholic hepatitis and alcoholic cirrhosis. J Pathol 186:82–89

Albano E (2006) Alcohol, oxidative stress and free radical damage. Proc Nutr Soc 65:278–290

Albano E (2008) Oxidative mechanisms in the pathogenesis of alcoholic liver disease. Mol Aspects Med 29:9–16

Aleffi S, Petrai I, Bertolani C et al. (2005) Upregulation of proinflammatory and proangiogenic cytokines by leptin in human hepatic stellate cells. Hepatology 42:1339–1348

Aviram M, Rosenblat M, Billecke S et al. (1999) Human serum paraoxonase (PON1) is inactivated by oxidized low density lipoprotein and preserved by antioxidants. Free Rad Biol Med 26:892–904

Aviram M, Hardak E, Vaya J et al. (2000) Human serum paraoxonases (PON1) Q and R selectively decrease lipid peroxides in human coronary and carotid atherosclerotic lesions. PON1 esterase and peroxidase-like activities. Circulation 101:2510–2517

Batey RG, Cao Q, Gould B (2002) Lymphocyte-mediated liver injury in alcohol-related hepatitis. Alcohol 27:37–41

Birjmohun RS, Van Leuven SI, Levels JHM et al. (2007) High-density lipoprotein attenuates inflammation and coagulation on endotoxin challenge in humans. Arterioscler Thromb Vasc Biol 27:1153–1158

Burlina A, Galzigna L (1974) Serum arylesterase isoenzymes in chronic hepatitis. Clin Biochem 7:202–205

Burlina A, Michielin E, Galzigna L (1977) Characteristics and behaviour of arylesterase in human serum and liver. Eur J Clin Inves 7:17–20

Cabré M, Ferré N, Folch J et al. (1999) Inhibition of hepatic cell nuclear DNA fragmentation by zinc in carbon tetrachloride-treated rats. J Hepatol 31:228–234

Calabresi L, Gomaraschi M, Vilia B et al. (2002) Elevated soluble adhesion molecules in subjects with low HDL-cholesterol. Arterioscler Thromb Vasc Biol 22:656–661

Camps J, Marsillach J, Joven J (2007) Measurement of serum paraoxonase-1 activity as a potential biomarker for chronic liver impairment. Clin Chim Acta 386:114–115

Chittury S, Farrell GC (2001) Ethiopathogenesis of nonalcoholic steatohepatitis. Semin Liver Dis 21:27–41

Choi J, Lee KJ, Zheng Y et al. (2004) Reactive oxygen species suppress hepatitis C virus RNA replication in human hepatoma cells. Hepatology 39:81–89

Choi J, Ou JHJ (2006) Mechanisms of liver injury. III. Oxidative stress in the pathogenesis of hepatitis C virus. Am J Physiol Gastrointest Liver Physiol 290:G847–G851

Clay MA, Pyle DH, Rye KA et al. (2001) Time sequence of inhibition of endothelial adhesion molecule expression by reconstituted high density lipoproteins. Atherosclerosis 157:23–29

Cockerill GW, Rye KA, Gamble JR et al. (1995) High density lipoproteins inhibit cytokine induced expression of endothelial cell adhesion molecules. Arterioscler Thromb Vasc Biol 15:1987–1994

Costa LG, Vitalone A, Cole TB et al. (2005) Modulation of paraoxonase (PON1) activity. Biochem Pharmacol 69:541–550

Crabb DW, Liangpunsakul S (2006) Alcohol and lipid metabolism. J Gastroenterol Hepatol 21:S56–S60

Day CP (2006) Genes or environment to determine alcoholic liver disease and non-alcoholic fatty liver disease. Liver Int 26:1021–1028

Deakin S, Leviev I, Gomaraschi M et al. (2002) Enzymatically active paraoxonase-1 is located at the external membrane of producing cells and released by a high affinity, saturable, desorption mechanism. J Biol Chem 277:4301–4308

Deakin S, Leviev I, Brulhart-Meynet MC et al. (2003) Paraoxonase-1 promoter haplotypes and serum paraoxonase: a predominant rele for polymorphic position –107, implicating the Sp1 transcription factor. Biochem J 372:643–649

Devalaraja MN, McClain CJ,Barve S et al. (1999) Increased monocyte MCP-1 production in acute alcoholic hepatitis. Cytokine 11:875–881

Ferré N, Camps J, Cabré M et al. (2001) Hepatic paraoxonase activity alterations and free radical production in rats with experimental cirrhosis. Metabolism 50:997–1000

Ferré N, Camps J, Prats E et al. (2002) Serum paraoxonase activity: a new additional test for the improved evaluation of chronic liver damage. Clin Chem 48:261–268

Ferré N, Marsillach J, Camps J et al. (2005) Genetic association of paraoxonase-1 polymorphisms and chronic hepatitis C virus infection. Clin Chim Acta 361:206–210

Ferré N, Marsillach J, Camps J et al. (2006). Paraoxonase-1 is associated with oxidative stress, fibrosis and FAS expression in chronic liver diseases. J Hepatol 45:51–59

Fisher NC, Neil DAH, Williams A et al. (1999) Serum concentrations and peripheral secretion of the beta chemokines monocyte chemoattractant protein 1and macrophage inflammatory protein 1α in alcoholic liver disease. Gut 45:426–430

Forman HJ, Torres M, Fukuto J (2003) Signal transduction by reactive oxygen and nitrogen species: pathways and chemical principles. Kluwer Academic, Boston

Fromenty B, Grimbert S, Mansouri A et al. (1995) Hepatic mitochondrial DNA deletion in alcoholics: association with microvesicular steatosis. Gastroenterology 108:193–200

Fürnsinn C, Willson TM, Brunmair B (2007) Peroxisome proliferator-activated receptor-δ, a regulator of oxidative capacity, fuel switching and cholesterol transport. Diabetologia 50:8–17

James RW, Deakin SP (2004) The importance of high-density lipoproteins for paraoxonase-1 secretion, stability, and activity. Free Rad Biol Med 37:1986–1994

Jiang Y, Beller DI, Frendl G et al. (1992) Monocyte chemoattractant protein-1regulates adhesion molecule expression and cytokine production in human monocytes J Immunol 148:2423–2428

Kawai H, Sakamoto F, Inoue Y (1990) Improved specific assay for serum arylesterase using a water-soluble substrate. Clin Chim Acta 188:177–182

Kharbanda KK, McVicker DL, Zetterman RK et al. (1995) Ethanol consumption reduces the proteolytic capacity and protease activities of hepatic lysosomes. Biochem Biophys Acta 1245:421–429

Kilic SS, Aydin S, Kilic N et al. (2005) Serum arylesterase and paraoxonase activity in patients with chronic hepatitis. World J Gastroenterol 11:7351–7354

Kleiner DE, Brunt EM, Natta MV et al. (2005) Design and validation of a histological scoring system for nonalcoholic fatty liver disease. Hepatology 41:1313–1321

Mackness B, Hine D, Liu Y et al. (2004) Paraoxonase-1 inhibits oxidised LDL-induced MCP-1 production by endothelial cells. Biochem Biophys Res Commun 318:680–683

Mahmood S, Kawanaka M, Kamei A et al. (2004) Immunohistochemical evaluation of oxidative stress markers in chronic hepatitis C. Antioxid Redox Signal 6:19–24

Marsillach J, Bertran N, Camps J et al. (2005) The role of circulating monocyte chemoattractant protein-1 as a marker of hepatic inflammation in patients with chronic liver disease. Clin Biochem 38:1138–1140

Marsillach J, Ferré N, Vila MC et al. (2007) Serum paraoxonase-1 in chronic alcoholics: Relationship with liver disease. Clin Biochem 40:645–650

Marsillach J, Aragonès G, Beltrán R et al. (2008) The measurement of the lactonase activity of paraoxonase-1 in the clinical evaluation of patients with chronic liver impairment. Clin Biochem doi:10.1016/j.clinbiochem.2008.09.120

Marsillach J, Camps J, Ferré N et al. (2009) Paraoxonase-1 is related to inflammation, fibrosis and PPARδ in experimental liver disease. BMC Gastroenterol 9:3

Mottaran E, Stewart SF, Rolla R et al. (2002) Lipid peroxidation contributes to immune reactions associated with alcoholic liver disease. Free Rad Biol Med 32:38–48

Nanji AA (2004) Role of different dietary fatty acids in the pathogenesis of experimental alcoholic liver disease. Alcohol 34:21–25

Navab M, Imes SS. Hama SY et al. (1991) Monocyte transmigration induced by modification of low density lipoprotein in cocultures of human aortic wall cells is due to induction of monocyte chemotactic protein 1 synthesis and is abolished by high density lipoprotein. J Clin Invest 88:2039–2046

Navab M, Hama SY, Cooke CJ et al. (2000a) Normal high density lipoprotein inhibits three steps in the formation of mildly oxidized low density lipoprotein: step 1. J Lip Res 41:1481–1484

Navab M, Hama SY, Anantharamaiah GM et al. (2000b) Normal high density lipoprotein inhibits three steps in the formation of mildly oxidized low density lipoprotein: steps 2 and 3. J Lip Res 41:1485–1508

Navab M, Berliner JA, Subbanagounder G et al. (2001) HDL and the inflammatory response induced by LDL-derived oxidised phospholipids. Arterioscler Thromb Vasc Biol 21:481–488

Navab M, Ananthramaiah GM, Reddy ST et al. (2004) The oxidation hypothesis of atherogenesis: the role of oxidized phospholipids and HDL. J Lip Res 45:993–1007.

Negre-Salvayre A, Dousset N, Ferretti G et al. (2006) Antioxidant and cytoprotective properties of high-density lipoproteins in vascular cells. Free Rad Biol Med 41:1031–1040

Nieto N (2007) Ethanol and fish oil induce NFkB transactivation of the collagen a2(1) promoter through lipid peroxidation-driven activation of PKC-PI3K-Akt pathway. Hepatology 45: 1433–1445

Nofer JR, Kehrel B, Fobker M et al. (2002) HDL and arteriosclerosis: beyond reverse cholesterol transport. Atherosclerosis 161:1–16

Pan M, Cederbaum AI, Zhang YL et al. (2004) Lipid peroxidation and oxidant stress regulate hepatic apolipoprotein B degradation and VLDL production. J Clin Invest 113:1277–1287

Parola M, Bellomo G, Robino G et al. (1999) 4-Hydroxynonenal as a biological signal: molecular basis and pathophysiological implications. Antioxidant Redox Signaling 1:255–284

Pesant M, Sueur S, Dutartre P et al. (2006) Peroxisome proliferator-activated receptor δ (PPARδ) activation protects H9c2 cardiomyoblasts from oxidative stress-induced apoptosis. Cardiovasc Res 69:440–449

Pessayre D, Fromenty B (2005) NASH a mitochondrial disease. J Hepatol 42:928–940

Rao MN, Marmillot P, Gong M et al. (2003) Light, but not heavy alcohol drinking, stimulates paraoxonase by upregulating liver mRNA in rats and humans. Metabolism 52:1287–1294

Simpson KJ, Henderson NC, Bone-Larson CL et al. (2003) Chemokines in the pathogenesis of liver disease: so many players with poorly defined roles. Clin Sci (Lond) 104:47–63

Solís Herruzo JA, García Ruíz I, Pérez Carreras M et al. (2006) Non-alcoholic fatty liver disease. from insulin resistance to mitochondrial dysfunction. Rev Esp Enferm Dig (Madrid) 98: 844–874

Sugano M, Tsuchida K, Makino N (2000) High-density lipoproteins protect endothelial cells from tumor necrosis factor-a-induced apoptosis. Biochem Biophys Res Commun 272: 872–876

Tamura Y, Sugimoto M, Muruyama T et al. (2008) Inhibition of CCR2 ameliorates insulin resistance and hepatic steatosis in db/db mice. Arterioscler Thromb Vasc Biol doi:10.1161/ATVBAHA.108.168633

Tardif KD, Waris G, Siddiqui A (2005) Hepatitis C virus, ER stress, and oxidative stress. Trends Microbiol 13:159–163

Thiele GM, Freeman TK, Klassen LW (2004) Immunological mechanisms of alcoholic liver disease. Semin Liver Dis 24:273–287

Thoren F, Romero A, Lindh M et al. (2004) A hepatitis C virus-encoded, nonstructural protein (NS3) triggers dysfunction and apoptosis in lymphocytes: role of NADPH oxidase-derived oxygen radicals. J Leukoc Biol 76:1180–1186

Tous M, Ferré N, Rull A et al (2006) Dietary cholesterol and differential monocyte chemoattractant protein-1 gene expression in aorta and liver of apo E-deficient mice Biochem Biophys Res Commun 340:1078–1084

Tsukamoto H, Lu SC (2001) Current concepts in the pathogenesis of alcoholic liver injury. FASEB J 15:1335–1349

Tsuruta S, Nakamuta M, Enjoji M et al. (2004) Anti-monocyte chemoattractant protein-1 gene therapy prevents dimethylnitrosamine-induced hepatic fibrosis in rats. Int J Mol Med 14: 837–842

Younossi ZM, Diehl AM, Ong JP (2002) Nonalcoholic fatty liver disease: An agenda for clinical research. Hepatology 35:746–752

Xu GY, Lv GC, Chen Y et al. (2005) Monitoring the level of serum paraoxonase 1 activity in liver transplantation patients. Hepatobiliary Pancreat Dis Int 4:178–181

Zamara E, Novo E, Marra F et al. (2004) 4-Hydroxynonenal as a selective pro-fibrogenic stimulus for activated human hepatic stellate cells. J Hepatol 40:60–68

Biomarkers of Sensitivity and Exposure in Washington State Pesticide Handlers

Jonathan N. Hofmann, Matthew C. Keifer, Harvey Checkoway, Anneclaire J. De Roos, Federico M. Farin, Richard A. Fenske, Rebecca J. Richter, Gerald van Belle, and Clement E. Furlong

Abstract Organophosphate (OP) and *N*-methyl-carbamate (CB) insecticides are widely used in agriculture in the US and abroad. These compounds – which inhibit acetylcholinestersase (AChE) enzyme activity – continue to be responsible for a high proportion of pesticide poisonings among US agricultural workers. It is possible that some individuals may be especially susceptible to health effects related to OP/CB exposure. The paraoxonase (PON1) enzyme metabolizes the highly toxic oxon forms of some OPs, and an individual's PON1 status may be an important determinant of his or her sensitivity to these chemicals. This chapter discusses methods used to characterize the PON1 status of individuals and reviews previous epidemiologic studies that have evaluated PON1-related sensitivity to OPs in relation to various health endpoints. It also describes an ongoing longitudinal study among OP-exposed agricultural pesticide handlers who are participating in a recently implemented cholinesterase monitoring program in Washington State. This study will evaluate handlers' PON1 status as a hypothesized determinant of butyrylcholinesterase (BuChE) inhibition. Such studies will be useful to determine how regulatory risk assessments might account for differences in PON1-related OP sensitivity when characterizing inter-individual variability in risk related to OP exposure. Recent work assessing newer and more sensitive biomarkers of OP exposure is also discussed briefly in this chapter.

Keywords Agriculture · Cholinesterase · Farm workers · Gene–environment interaction · Organophosphates · Paraoxonase (PON1) · Pesticides

1 Introduction

Since the 1970s, the use of organophosphate (OP) and *N*-methyl-carbamate (CB) insecticides has increased dramatically in the US (Reigart and Roberts, 1999). In 2007, approximately 589,000 lbs of azinphos-methyl, chlorpyrifos, and carbaryl

J.N. Hofmann (✉)
Division of Cancer Epidemilogy and Genetics, National Cancer Institute, Bethesda, MD, USA
e-mail: hofmannjn@mail.nih.gov

S.T. Reddy (ed.), *Paraoxonases in Inflammation, Infection, and Toxicology*, Advances in Experimental Medicine and Biology 660, DOI 10.1007/978-1-60761-350-3_3, © Humana Press, a part of Springer Science+Business Media, LLC 2010

(three common OP/CB insecticides) were applied in apple orchards in Washington State alone (WASS, 2007). OP/CBs are also widely used to treat other crops including pears, cherries, grapes, hops, and potatoes (WSUCE, 2001; WASS, 2008; Smith, 2006).

Acute effects of OP/CB exposure have been well documented; inhibition of neuronal acetylcholinesterase (AChE) enzyme activity is the main mechanism of OP/CB toxicity (Ecobichon, 2001). AChE hydrolyzes the neurotransmitter acetyl-choline, and thereby plays a critical role in regulating nerve transmissions in the central and peripheral nervous systems (Ecobichon, 2001). Cholinesterases (ChE) are also found in blood in two different forms; AChE is associated with red blood cell membranes, and butyrylcholinesterase (BuChE) is present in serum (Wilson, 2001). Both AChE and BuChE inhibition are considered to be surrogate markers of early biologic effects related to OP/CB exposure (USEPA, 2000). Generally, AChE inhibition is considered to be a better marker of toxicity, whereas BuChE inhibition is a more sensitive marker of exposure because it is inhibited more effectively than AChE by most OP/CBs including chlorpyrifos, diazinon, and malathion (Lotti, 1995, 2001; Yuknavage et al. 1997).

2 ChE Monitoring in Washington State

In 2004, the Washington State Department of Labor and Industries (L&I) initiated a ChE monitoring program for agricultural pesticide handlers who are exposed to toxicity class I or II OP/CBs (WAC, 2004). In this program, handlers (e.g., agricultural workers who mix, load, or apply pesticides) are tested for AChE and BuChE activities annually prior to the OP/CB spray season (i.e., at baseline), and during the spray season if they are exposed for 30 or more hours in a 30-day period. AChE or BuChE inhibition from baseline levels can lead to work practice evaluations or removal from continued OP/CB exposure (with wage protection) depending on the degree of ChE inhibition observed. The goals of this monitoring program are to identify and correct unsafe work practices, and to prevent further exposure among handlers with ChE inhibition before they experience symptoms of pesticide-related illness.

3 Health Effects of OP/CB Exposure

Among agricultural workers in the US, OP/CBs continue to be responsible for a high proportion of pesticide poisonings due to their high toxicity and widespread use in agriculture (Reigart and Roberts, 1999). In an analysis of acute pesticide poisonings among US agricultural workers from 1998 to 2005, Calvert et al. noted that OP/CBs were implicated more frequently than any other class of pesticides (Calvert et al. 2008). In addition to acute poisonings, there is also growing concern about a variety of health endpoints that may be associated with chronic exposure to OP/CB insecticides (McCauley et al. 2006). In particular, there is some evidence of associations between OP exposure and chronic neurologic effects (Kamel et al.

2005; Kamel and Hoppin, 2004; Rothlein et al. 2006) and various types of cancer (Alavanja et al. 2004).

4 Sensitivity to OP/CB Toxicity

It is possible that some individuals may be especially susceptible to health effects related to OP/CB exposure. High density lipoprotein (HDL)-associated paraoxonase (PON1) is thought to be one important determinant of an individual's sensitivity to some OP insecticides, based primarily on evidence from studies in animal models (Cole et al. 2005; Li et al. 2000; Shih et al. 1998). PON1 hydrolyzes the highly toxic oxon forms of several widely used OPs, including chlorpyrifos and diazinon. Studies in transgenic mice have clearly demonstrated that low plasma PON1 activity is associated with greater brain AChE inhibition following exposure to chlorpyrifos oxon and diazoxon (the oxon forms of chlorpyrifos and diazinon) (Li et al. 2000). Also, a Q/R polymorphism at position 192 in the *PON1* coding region affects the catalytic efficiency of the enzyme for chlorpyrifos oxon metabolism. In a study by Cole et al. of mice expressing equivalent levels of the different alloforms of humanized PON1, greater brain AChE inhibition was observed among mice with the Q alloform relative to the R alloform following chlorpyrifos oxon exposure (Cole et al. 2005).

Several important points should be considered regarding PON1-mediated sensitivity to OP exposure: (1) PON1 status is most relevant for protecting against direct exposure to the oxon forms of OP insecticides (Li et al. 2000; Shih et al. 1998); (2) most – if not all – OP exposures include oxon residues (CalEPA, 1996; Yuknavage et al. 1997); and (3) the safety studies for OPs such as chlorpyrifos were carried out with the highly pure parent organophosphorothioate (Nolan et al. 1984).

5 Characterizing PON1 Status

An individual's functional PON1 Q192R genotype can be determined using a two-substrate enzymatic analysis. In this analysis, paraoxonase (POase) and diazoxonase (DZOase) activities are measured in plasma samples, and the results are plotted on a graph (Fig. 1). Methods for these assays and the two-substrate enzymatic analysis have been described previously (Furlong et al. 1989; Jarvik et al. 2000; Richter and Furlong, 1999). Previous studies have found that there is generally excellent agreement between predicted Q192R genotype determined by the two-substrate analysis and observed genotype using polymerase chain reaction (PCR) assays (Jarvik et al. 2003).

In addition to assessing functional Q192R genotype, it is also useful to characterize the level of plasma PON1 activity by measuring arylesterase (AREase) activity in plasma samples using phenylacetate as the substrate. AREase activity is considered to be a good surrogate for PON1 concentration in plasma since its rate of hydrolysis is not affected by the Q192R polymorphism (Furlong et al. 1993, 2006).

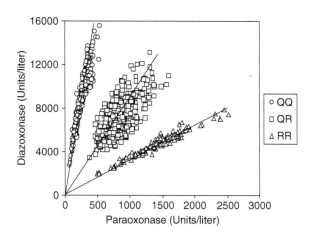

Fig. 1 Example of the two-substrate enzyme activity distribution plot for determination of PON1 status [figure reproduced from Richter et al., 2004 with permission]

6 Previous Epidemiologic Studies

Despite convincing evidence in animal models, relatively few epidemiologic studies have evaluated PON1 status as a determinant of OP sensitivity. Mackness et al. conducted a case–control study of self-reported chronic ill health among sheep dippers exposed primarily to diazinon (Mackness et al. 2003). Cases were individuals who believed that their chronic ill health was due to exposure to sheep dip, and controls were sheep dippers who were believed to be in good health. The investigators found that farmers in the lowest quintile of DZOase activity had a 2.5-fold higher risk of chronic ill health. Another study by Lee et al. evaluated *PON1* Q192R genotype among OP-exposed fruit farm workers in South Africa (Lee et al. 2003). Relative to RR individuals, those who were either QR heterozygotes or QQ homozygotes were almost three times as likely to report multiple (≥ 2) symptoms of chronic OP toxicity (e.g., abdominal pain, headache, gait disturbance, and limb numbness, among other symptoms). In a case–control study of acute OP intoxication, Sozmen et al. (2002) found that cases had a significantly higher frequency of the *PON1* Q192 alloform and lower POase activity than controls. They also found that POase activity was lower among cases with low BuChE activity upon hospital admission relative to cases with higher BuChE activity, suggesting a protective effect of PON1 against BuChE inhibition.

7 PON1 Status and BuChE Inhibition Among Pesticide Handlers

In 2006, we began a study to evaluate BuChE inhibition in relation to PON1 status among OP-exposed agricultural pesticide handlers in Washington State. Pesticide handlers in the statewide ChE monitoring program are recruited through two collaborating clinics in eastern Washington during the OP/CB spray season (April–July).

Blood samples for determination of PON1 status and self-reported OP/CB exposure information are collected at the time of follow-up ChE testing. Data from the 2006 and 2007 spray seasons have been analyzed, and results have recently been published (Hofmann et al. 2009). We focused on the outcome of BuChE inhibition in this analysis for several reasons. First, BuChE is more sensitive than AChE to inhibition by many OPs including chlorpyrifos, which was the most widely used OP among study participants (Amitai et al. 1998; Lotti, 2001). Second, there was little evidence of substantial AChE inhibition among study participants or among all handlers in the statewide monitoring program. Among the 472 handlers in the state monitoring program who had baseline tests and at least one follow-up test in 2006, mean AChE inhibition was 1.8%, and only two handlers had "AChE depression" at the work practice evaluation threshold of >20% inhibition (ChESAC, 2006). Finally, high variability in AChE measurements was observed in analyses of data from the state monitoring program in 2007 (16.7% CV) (Kalman and van Belle, 2007); this would likely have obscured any associations between PON1 status and AChE inhibition in our study.

Our study approach has several strengths. By recruiting participants from the recently implemented Washington State ChE monitoring program, we have established a cohort of agricultural pesticide handlers with confirmed recent OP exposure. Because pesticide handlers (i.e., mixer/loader/applicators) are considered to be more highly exposed to pesticides than agricultural workers who perform other activities, this population is ideal for evaluating PON1-mediated susceptibility to OP exposure. Previous studies of PON1-related susceptibility among individuals with occupational OP exposure have relied on self-reported health outcomes such as chronic ill health (Mackness et al. 2003) or symptoms of chronic toxicity (Lee et al. 2003). In this study, we used BuChE inhibition as a quantitative biomarker of OP-related effects as our primary outcome. Finally, some previous studies have relied exclusively on *PON1* genotype. For some OPs (i.e., chlorpyrifos), the Q192R polymorphism is important as it affects the catalytic efficiency of hydrolysis. However, plasma PON1 activity level is important for all OPs that are metabolized by PON1 at physiologically relevant rates (Li et al. 2000). Plasma PON1 activity has been shown to vary widely among individuals with the same Q192R genotype (Costa et al. 2005; Eckerson et al. 1983; Furlong et al. 2006). Determination of both *PON1* genotype and level of plasma PON1 activity in this study allows for a better characterization of overall PON1 status than Q192R genotype alone (Richter and Furlong, 1999).

However, there are also several limitations to our study methods. The cross-sectional design limits our ability to infer a causal relationship between plasma PON1 activity and BuChE inhibition, because PON1 activity may be modified to some extent by certain medications (e.g., statins), dietary habits (e.g., vitamin C and E intake), and environmental exposures (e.g., smoking) (Durrington et al. 2002; Jarvik et al. 2002). However, most previous studies suggest that plasma PON1 activity tends to be relatively stable over time and is mostly regulated by genetic factors, particularly the C-108T promoter polymorphism (Brophy et al. 2001a, b; Durrington et al. 2002; Ferre et al. 2003; Furlong et al. 2000; Jarvik et al. 2002; Leviev and James, 2000; Suehiro et al. 2000; Zech and Zurcher, 1974). Nonetheless, it is possible that some exposures that modulate plasma PON1 activity may also affect BuChE activity or recovery following OP/CB exposure. Future studies with prospective collection of blood specimens for determination of PON1 status are needed.

Other studies may also benefit from recent work evaluating new biomarkers of OP exposure. Several previous studies have found that acylpeptide hydrolase (APH) is particularly sensitive to inhibition by some OPs, in that biological effects can be observed at levels of OP exposure that do not inhibit traditional biomarkers such as AChE (Quistad et al. 2005; Richards et al. 1999). Moreover, there are several other advantages of using APH as a biomarker of OP exposure. Whereas BuChE is relatively short-lived in serum, APH is present in red blood cells and therefore has a longer life span. Consequently, it may be possible to detect exposures that occurred during the preceding several months. Using mass spectrometry (MS), it should be possible in some cases to identify which particular OP compound resulted in APH inhibition. MS analyses should also provide a more accurate measure of exposure when used with appropriate heavy isotope-labeled internal standard biomarker proteins spiked into the starting samples.

8 Conclusions

Based on evidence from extensive research in animal models and some epidemiologic studies, regulatory risk assessments should take differences in PON1-related sensitivity to OP insecticides into consideration when characterizing inter-individual variability in risk related to OP exposure. At some point in the future, biologic monitoring for PON1 status among pesticide handlers may be warranted to identify individuals who are at particularly high risk of OP-related health effects. However, issues of test validity as well as the ethical and legal aspects of genetic testing in the workplace need to be addressed before such a program could be implemented (Battuello et al. 2004).

Acknowledgments We would like to thank all of the workers who participated in this study. We also acknowledge the following individuals who contributed to this study: Zahra Afsharinejad for her work on the PON1 genotyping assays; Kelly Fryer-Edwards for her assistance with outreach to study participants; Pam Ernst and Joe Cozzetto from Central Washington Occupational Medicine for their assistance with our recruitment efforts; and Maria Negrete and Pablo Palmandez from the Pacific Northwest Agricultural Safety and Health Center for their assistance with field data collection. Financial support for this project was provided by CDC/NIOSH grants U50OH07544 and T42OH008433, and NIEHS grants P30ES07033, T32ES07262, P42ES04696, and ES009883.

References

Alavanja MC, Hoppin JA, Kamel F. Health effects of chronic pesticide exposure: cancer and neurotoxicity. Annu Rev Public Health 2004;25:155–197.
Amitai G, Moorad D, Adani R, Doctor BP. Inhibition of acetylcholinesterase and butyrylcholinesterase by chlorpyrifos-oxon. Biochem Pharmacol 1998;56(3):293–299.

Battuello K, Furlong CE, Fenske R, Austin MA, Burke W. Paraoxonase polymorphisms and susceptibility to organophosphate pesticides. In: Khoury MJ, Little J, Burke W (eds.). Human Genome Epidemiology: A Scientific Foundation for Using Genetic Information to Improve Health and Prevent Disease. New York: Oxford University Press; 2004.

Brophy VH, Hastings MD, Clendenning JB, Richter RJ, Jarvik GP, Furlong CE. Polymorphisms in the human paraoxonase (PON1) promoter. Pharmacogenetics 2001;11(1):77–84.

Brophy VH, Jampsa RL, Clendenning JB, McKinstry LA, Jarvik GP, Furlong CE. Effects of 5′ regulatory-region polymorphisms on paraoxonase-gene (PON1) expression. Am J Hum Genet 2001;68(6):1428–1436.

CalEPA. Report for the application and ambient air monitoring of chlorpyrifos (and the oxon analogue) in Tulare County during spring/summer, 1996. Sacramento, CA: California Environmental Protection Agency; 1998.

Calvert GM, Karnik J, Mehler L, Beckman J, Morrissey B, Sievert J, et al. Acute pesticide poisoning among agricultural workers in the United States, 1998–2005. Am J Ind Med 2008;51(12):883–898

ChESAC. Cholinesterase Monitoring of Pesticide Handlers in Agriculture: 2004–2006. Olympia, WA: Scientific Advisory Committee for Cholinesterase Monitoring; 2006 November 13.

Cholinesterase Monitoring. In: Washington Administrative Code, Chapter 296–307, Part J-1; 2004.

Cole TB, Walter BJ, Shih DM, Tward AD, Lusis AJ, Timchalk C, et al. Toxicity of chlorpyrifos and chlorpyrifos oxon in a transgenic mouse model of the human paraoxonase (PON1) Q192R polymorphism. Pharmacogenet Genomics 2005;15(8):589–598.

Costa LG, Cole TB, Vitalone A, Furlong CE. Measurement of paraoxonase (PON1) status as a potential biomarker of susceptibility to organophosphate toxicity. Clin Chim Acta 2005;352 (1–2):37–47.

Durrington PN, Mackness B, Mackness MI. The hunt for nutritional and pharmacological modulators of paraoxonase. Arterioscler Thromb Vasc Biol 2002;22(8):1248–1250.

Eckerson HW, Wyte CM, La Du BN. The human serum paraoxonase/arylesterase polymorphism. Am J Hum Genet 1983;35(6):1126–1138.

Ecobichon DJ. Toxic effects of pesticides. In: Klaassen CD (ed.). Casarett and Doull's Toxicology: The Basic Science of Poisons. 6th ed. New York: McGraw-Hill; 2001. 763–810.

Ferre N, Camps J, Fernandez-Ballart J, Arija V, Murphy MM, Ceruelo S, et al. Regulation of serum paraoxonase activity by genetic, nutritional, and lifestyle factors in the general population. Clin Chem 2003;49(9):1491–1497.

Furlong CE, Costa LG, Hassett C, Richter RJ, Sundstrom JA, Adler DA, et al. Human and rabbit paraoxonases: purification, cloning, sequencing, mapping and role of polymorphism in organophosphate detoxification. Chemico-Biological Interactions 1993;87(1–3): 35–48.

Furlong CE, Holland N, Richter RJ, Bradman A, Ho A, Eskenazi B. PON1 status of farmworker mothers and children as a predictor of organophosphate sensitivity. Pharmacogenetics and Genomics 2006;16(3):183–190.

Furlong CE, Li WF, Richter RJ, Shih DM, Lusis AJ, Alleva E, et al. Genetic and temporal determinants of pesticide sensitivity: role of paraoxonase (PON1). Neurotoxicology 2000;21(1–2):91–100.

Furlong CE, Richter RJ, Seidel SL, Costa LG, Motulsky AG. Spectrophotometric assays for the enzymatic hydrolysis of the active metabolites of chlorpyrifos and parathion by plasma paraoxonase/arylesterase. Anal Biochem 1989;180(2):242–247.

Hofmann JN, Keifer MC, Furlong CE, De Roos AJ, Farin FM, Fenske RA, et al. Serum cholinesterase inhibition in relation to paraoxonase-1 (PON1) status among organophosphate-exposed agricultural pesticide handlers. Environ Health Perspect 2009;117(9)1402–1408.

Jarvik GP, Jampsa R, Richter RJ, Carlson CS, Rieder MJ, Nickerson DA, et al. Novel paraoxonase (PON1) nonsense and missense mutations predicted by functional genomic assay of PON1 status. Pharmacogenetics 2003;13(5):291–295.

Jarvik GP, Rozek LS, Brophy VH, Hatsukami TS, Richter RJ, Schellenberg GD, et al. Paraoxonase (PON1) phenotype is a better predictor of vascular disease than is PON1(192) or PON1(55) genotype. Arterioscler Thromb Vasc Biol 2000;20(11):2441–2447.

Jarvik GP, Tsai NT, McKinstry LA, Wani R, Brophy VH, Richter RJ, et al. Vitamin C and E intake is associated with increased paraoxonase activity. Arterioscler Thromb Vasc Biol 2002;22(8):1329–1333.

Kalman D, van Belle G. 2007 Cholinesterase Analysis. URL: http://www.lni.wa.gov/ Safety/Topics/AtoZ/Cholinesterase/files/report120307.pdf. Letter. Seattle, WA: University of Washington; 2007.

Kamel F, Engel LS, Gladen BC, Hoppin JA, Alavanja MC, Sandler DP. Neurologic symptoms in licensed private pesticide applicators in the agricultural health study. Environ Health Perspect 2005;113(7):877–882.

Kamel F, Hoppin JA. Association of pesticide exposure with neurologic dysfunction and disease. Environ Health Perspect 2004;112(9):950–958.

Lee BW, London L, Paulauskis J, Myers J, Christiani DC. Association between human paraoxonase gene polymorphism and chronic symptoms in pesticide-exposed workers. J Occup Environ Med 2003;45(2):118–122.

Leviev I, James RW. Promoter polymorphisms of human paraoxonase PON1 gene and serum paraoxonase activities and concentrations. Arterioscler Thromb Vasc Biol 2000;20(2):516–521.

Li WF, Costa LG, Richter RJ, Hagen T, Shih DM, Tward A, et al. Catalytic efficiency determines the in-vivo efficacy of PON1 for detoxifying organophosphorus compounds. Pharmacogenetics 2000;10(9):767–779.

Lotti M. Cholinesterase inhibition: complexities in interpretation. Clin Chem 1995;41(12 Pt 2):1814–1818.

Lotti M. Clinical toxicology of anticholinesterase agents in humans. In: Krieger R (ed.). Handbook of Pesticide Toxicology. 2nd ed. San Diego, CA: Academic Press; 2001. 1043–1085.

Mackness B, Durrington P, Povey A, Thomson S, Dippnall M, Mackness M, et al. Paraoxonase and susceptibility to organophosphorus poisoning in farmers dipping sheep. Pharmacogenetics 2003;13(2):81–88.

McCauley LA, Anger WK, Keifer M, Langley R, Robson MG, Rohlman D. Studying health outcomes in farmworker populations exposed to pesticides. Environ Health Perspect 2006;114(6):953–960.

Nolan RJ, Rick DL, Freshour NL, Saunders JH. Chlorpyrifos: pharmacokinetics in human volunteers. Toxicol Appl Pharmacol 1984;73(1):8–15.

Quistad GB, Klintenberg R, Casida JE. Blood acylpeptide hydrolase activity is a sensitive marker for exposure to some organophosphate toxicants. Toxicol Sci 2005;86(2):291–299.

Reigart J, Roberts J. Recognition and Management of Pesticide Poisonings. 5th ed. Washington, DC: Environmental Protection Agency; 1999.

Richards P, Johnson M, Ray D, Walker C. Novel protein targets for organophosphorus compounds. Chem Biol Interact 1999;119–120:503–511.

Richter RJ, Furlong CE. Determination of paraoxonase (PON1) status requires more than genotyping. Pharmacogenetics 1999;9(6):745–753.

Richter RJ, Jampsa R, Jarvik GP, Costa L, Furlong CE. Determination of paraoxonase 1 (PON1) status and genotypes at specific polymorphic sites. In: Mains M, Costa L, Reed D, Hodgson E (ed.). Current Protocols in Toxicology. New York, NY: John Wiley and Sons; 2004.

Rothlein J, Rohlman D, Lasarev M, Phillips J, Muniz J, McCauley L. Organophosphate pesticide exposure and neurobehavioral performance in agricultural and non-agricultural Hispanic workers. Environ Health Perspect 2006;114(5):691–696.

Shih DM, Gu L, Xia YR, Navab M, Li WF, Hama S, et al. Mice lacking serum paraoxonase are susceptible to organophosphate toxicity and atherosclerosis. Nature 1998;394(6690):284–287.

Smith T. 2006 Crop Protection Guide for Tree Fruits in Washington. Pullman, WA: Washington State University Extension; 2006. Report No.: EB0419.

Sozmen EY, Mackness B, Sozmen B, Durrington P, Girgin FK, Aslan L, et al. Effect of organophosphate intoxication on human serum paraoxonase. Hum Exp Toxicol 2002;21(5):247–252.

Suehiro T, Nakamura T, Inoue M, Shiinoki T, Ikeda Y, Kumon Y, et al. A polymorphism upstream from the human paraoxonase (PON1) gene and its association with PON1 expression. Atherosclerosis 2000;150(2):295–298.

WASS. Washington's agricultural fruit chemical usage, 2007: Apples. Olympia, WA: Washington Agricultural Statistics Service; 2008.

Wilson B. Cholinesterases. In: Krieger RI (ed.). Handbook of Pesticide Toxicology. San Diego: Academic Press; 2001. 967–985.

Yuknavage KL, Fenske RA, Kalman DA, Keifer MC, Furlong CE. Simulated dermal contamination with capillary samples and field cholinesterase biomonitoring. J Toxicol Environ Health 1997;51(1):35–55.

Zech R, Zurcher K. Organophosphate splitting serum enzymes in different mammals. Comp Biochem Physiol B 1974;48(3):427–433.

WSUCE. Crop Profile for Hops in Washington. Richland, WA: Washington State University Cooperative Extension; 2001. Report No.: MISC0353E.

WASS. Washington Ag Chemical Use. URL: http://www.nass.usda.gov/Statistics_by_State/Washington/Publications/Ag_Chemical_Use/index.asp. In. Olympia, WA: Washington Agricultural Statistics Service; Accessed December 2008.

USEPA. The use of data on cholinesterase inhibition for risk assessments of organophosphorous and carbamate pesticides. Washington, DC: Office of Pesticide Programs; 2000.

Paraoxonase 1 Status as a Risk Factor for Disease or Exposure

Rebecca J. Richter, Gail P. Jarvik, and Clement E. Furlong

Abstract Human paraoxonase 1 (PON1) has broad substrate specificity and has been shown to protect against exposure to some organophosphorus (OP) insecticides due to its ability to hydrolyze toxic metabolites of some organophosphorothioate insecticides. PON1 status has been shown to be important in protecting against vascular disease, presumably due to the not-as-yet fully characterized role of the three PON proteins in modulating oxidative stress. More recently, all three PONs (1, 2, and 3) have been shown to inactivate the quorum sensing factor N-(3-oxododecanoyl)-L-homoserine lactone (3OC12-HSL) of *Pseudomonas*. Expression of human PON1 in *Drosophila* demonstrated the importance of PON1 in resistance to *Pseudomonas* infection. Many studies have examined only DNA single nucleotide polymorphisms as possible risk factors for disease or exposures. For all of the known functions of PON1, the level of PON1 enzyme is important and, in some cases, also the Q192R polymorphism. A simple high throughput two-substrate assay/analysis, plotting rates of diazoxon hydrolysis vs. paraoxon hydrolysis, provided both PON1 levels and functional Q192R phenotype/genotype. We have developed a new two-substrate assay/analysis protocol that provides PON1 status without use of toxic OP substrates. Factors were determined for inter-converting rates of hydrolysis of different substrates.

Keywords PON1 status · Paraoxonase · Diazoxon · Diazinon · Chlorpyrifos · Chlorpyrifos oxon · Carotid artery disease · Quorum sensing factor · OP exposure

1 Introduction

Human paraoxonase 1 (PON1) has broad substrate specificity and has been shown to protect against exposure to some organophosphorus (OP) insecticides due to its ability to hydrolyze their toxic oxon metabolites at physiologically relevant rates (Costa

R.J. Richter (✉)
Departments of Medicine (Division of Medical Genetics) and Genome Sciences, University of Washington, Seattle, WA, 98195, USA
e-mail: rrichter@u.washington.edu

S.T. Reddy (ed.), *Paraoxonases in Inflammation, Infection, and Toxicology*, Advances in Experimental Medicine and Biology 660, DOI 10.1007/978-1-60761-350-3_4,
© Humana Press, a part of Springer Science+Business Media, LLC 2010

et al. 1990; Li et al. 1993, 1995, 2000; Shih et al. 1998; Cole et al. 2005). The hydrolysis of other OPs, such as paraoxon (PO), while detectable with in vitro assays, is not physiologically relevant due to insufficient catalytic efficiency of hydrolysis (Li et al. 2000). The PON1 Q192R polymorphism affects the catalytic efficiency of hydrolysis of some PON1 substrates (Davies et al. 1996; Li et al. 2000). We introduced the term PON1 status to include both the functional $PON1_{192}$ genotype as well as the plasma level of PON1, both of which can be important in determining risk of disease or exposure (Li et al. 1993). In all cases, rates of detoxication of both endogenous and xenobiotic substrates are determined by the plasma level of PON1, provided that the catalytic efficiency of hydrolysis is physiologically significant. For some cases, such as the detoxication of chlorpyrifos oxon (CPO), plasma PON1 level and the Q192R polymorphism are both important, with the $PON1_{R192}$ alloform detoxifying CPO more efficiently than $PON1_{Q192}$ (Li et al. 2000). The efficiency of detoxication of diazoxon (DZO) is nearly equivalent for both $PON1_{192}$ alloforms (Li et al. 2000).

PON1 status has been shown to be important in protecting against vascular disease (Jarvik et al. 2000), presumably through the role of PON1 in modulating oxidative stress (reviewed in James 2006). More recently, all three PONs (1, 2, and 3) have been shown to inactivate the quorum sensing factor N-(3-oxododecanoyl)-L-homoserine lactone (3OC12-HSL) of *Pseudomonas* (Ozer et al. 2005). The definitive study by Stoltz et al. (2008; Chapter 17 in this book), where the expression of human PON1 in transgenic *Drosophila* resulted in increased resistance to infection by *Pseudomonas aeruginosa*, indicates that the PON family of proteins can also be considered as part of the innate immunity system (Chun et al. 2004).

After the genetic variability of PON1 was linked to cardiovascular disease, many studies have been carried out that examined only DNA single nucleotide polymorphisms (SNPs) as possible risk factors for disease or exposures. Some studies examined only the Q192R polymorphism, others both the Q192R and L55M polymorphisms and yet others have included the analysis of one or more promoter region polymorphisms, the most important of which appears to be the C-108T polymorphism that occurs in an Sp1 binding site (Deakin et al. 2003). Relatively few studies have examined the relationship between PON1 levels or activity and risk for disease. Mackness et al. (2001) found decreased levels of plasma PON1 (by ELISA) and paraoxonase (POase) activity among patients with coronary heart disease (CHD). They also carried out a meta-analysis of 18 previous studies, only three of which determined PON1 levels or activity. The assay of POase is not a good measure of risk for disease, since this activity is dramatically affected by the Q192R polymorphism with $PON1_{R192}$ having much higher POase activity than $PON1_{Q192}$. Since the gene frequencies for $PON1_{Q192}$ and $PON1_{R192}$ vary significantly among different ethnic groups, a mixture of individuals of different ethnic origin can significantly skew the data (Brophy et al. 2002). Mackness et al. (2001) recommended that "We, along with other authors, would strongly suggest that all further epidemiological studies into the role of PON1 and disease should include a measurement of the enzyme itself in addition to the genetic polymorphisms." This same recommendation was echoed by two other experienced PON1 research teams (Deakin and James 2004; La Du 2003).

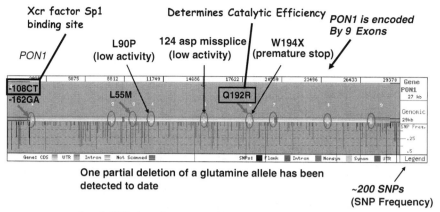

Fig. 1 The human *PON1* gene with known polymorphisms and their frequencies. The 5′ end of the gene is on the *left*. (Seattle SNPs, http://pga.gs.washington.edu/)

A second large study (Lawlor et al. 2004) examined association of the Q192R polymorphism with CHD in a large cohort ($n = 3,266$) combined with a meta-analysis of 38 other studies. These SNP analyses revealed no association with CHD; however, they suffered from the critical lack of data on plasma PON1 levels.

A third meta-analysis which examined four PON1 polymorphisms and one PON2 polymorphism included 43 genetic association studies (>11,000 cases and ~13,000 controls) and showed no significant association with CHD (Wheeler et al. 2004). This study also suffered from a lack of data on plasma PON1 levels or activity.

For all of the known functions of PON1, the level of PON1 is important and, in some cases, also the Q192R polymorphism. Figure 1 shows the PON1 polymorphisms and frequencies identified by the Seattle SNPs resequencing effort (Furlong et al. 2008). Characterizing all of the nearly 200 PON1 polymorphisms will not provide an accurate prediction of plasma PON1 levels. Measurement of the of the phenyl acetate hydrolysis activity of PON1 (AREase) is unaffected by the Q192R polymorphism and can serve as a surrogate measure of plasma PON1 protein levels (Furlong et al. 2006; Richter et al. 2008). Measurement of the AREase activity of plasma PON1 or determination of plasma PON1 protein levels by ELISA are the minimum measures that should be carried out in any epidemiological study. More useful measures are described below.

2 Two-Substrate Analyses of PON1 Status

Early studies by Eckerson et al. (1983) showed that a two-substrate assay/analysis plotting rates of PO hydrolysis vs. AREase would provide both plasma PON1 levels and a separation of low ($PON1_{Q192}$ homozygotes) from the high metabolizers. This analysis, however, did not resolve PON1$_{192}$ heterozygotes from PON1$_{R192}$ homozygotes. When we made use of the two-substrate assay/analysis to examine rates of a number of different PON1 substrates, we found that plotting rates of diazoxon hydrolysis vs. paraoxon hydrolysis provided both relative PON1 levels for each PON1$_{192}$ functional genotype/phenotype as well as a clear resolution of all three PON1$_{192}$ phenotypes (Q/Q, Q/R, and R/R) (Davies et al. 1996; Richter and Furlong 1999). This analysis, however, uses two highly toxic OPs, DZO and PO.

3 Development of a PON1 Status Protocol with Non-OP Substrates

Since PON1 status appears to be an important risk factor for OP exposure as well as for a number of diseases, we examined rates of hydrolysis of more than 70 substrates under different conditions of salt concentration and pH to develop a high throughput PON1 status protocol that did not make use of the highly toxic OP substrates. Figure 2 shows a comparison of the DZOase vs. POase protocol for

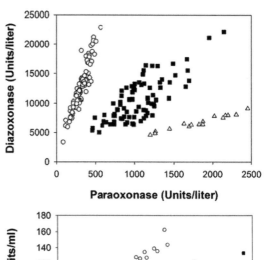

Fig. 2 Comparison of the two protocols for determining PON1 status. (**a**) Assays using the highly toxic OP substrates DZO and PO; and (**b**) assays using the non-OP substrates phenyl acetate and CMPA. The 183 plasma samples included 86 PON1$_{Q192}$ homozygotes, 79 heterozygotes and 18 PON1$_{R192}$ homozygotes with genotypes verified by PCR. Reproduced from Richter et al. (2009) with permission

Table 1 Conversion factors for rates of substrate hydrolysis

Phenotype	Conversion factors	$r^{2\,a}$
QQ	$AREase_{HS}{}^{b}$ (U/ml) $\times$ 172 = $DZOase_{phys}$ (U/L)^{c}	0.93
QR	$AREase_{HS}$ (U/ml) $\times$ 204 = $DZOase_{phys}$ (U/L)	0.82
RR	$AREase_{HS}$ (U/ml) $\times$ 286 = $DZOase_{phys}$ (U/L)	0.87
QQ	$AREase_{HS}$ (U/ml) $\times$ 69 = $CPOase_{phys}{}^{d}$ (U/L)	0.87
QR	$AREase_{HS}$ (U/ml) $\times$ 103 = $CPOase_{phys}$ (U/L)	0.88
RR	$AREase_{HS}$ (U/ml) $\times$ 189 = $CPOase_{phys}$ (U/L)	0.89
QQ	$AREase_{LS}{}^{e}$ (U/ml) $\times$ 110 = $DZOase_{phys}$ (U/L)	0.84
QR	$AREase_{LS}$ (U/ml) $\times$ 100 = $DZOase_{phys}$ (U/L)	0.72
RR	$AREase_{LS}$ (U/ml) $\times$ 83 = $DZOase_{phys}$ (U/L)	0.93
QQ	$AREase_{LS}$ (U/ml) $\times$ 45 = $CPOase_{phys}$ (U/L)	0.73
QR	$AREase_{LS}$ (U/ml) $\times$ 50 = $CPOase_{phys}$ (U/L)	0.84
RR	$AREase_{LS}$ (U/ml) $\times$ 55 = $CPOase_{phys}$ (U/L)	0.92
QQ	$AREase_{HS}$ (U/ml) $\times$ 3.8 = POase (U/L)	0.75
QR	$AREase_{HS}$ (U/ml) $\times$ 15.9 = POase (U/L)	0.50
RR	$AREase_{HS}$ (U/ml) $\times$ 47.6 = POase (U/L)	0.90
QQ[f]	$AREase_{HS}$ (U/ml) $\times$ 1.6 = $AREase_{LS}$ (U/ml)	0.85
QR[f]	$AREase_{HS}$ (U/ml) $\times$ 2.0 = $AREase_{LS}$ (U/ml)	0.66
RR[f]	$AREase_{HS}$ (U/ml) $\times$ 3.5 = $AREase_{LS}$ (U/ml)	0.83
QQ	$DZOase_{phys}$ (U/L) $\times$ 1.08 = $DZOase_{HS}{}^{g}$ (U/L)	0.90
QR	$DZOase_{phys}$ (U/L) $\times$ 1.01 = $DZOase_{HS}$ (U/L)	0.91
RR	$DZOase_{phys}$ (U/L) $\times$ 0.84 = $DZOase_{HS}$ (U/L)	0.87

[a]Correlation coefficient squared
[b]$AREase_{HS}$ = Arylesterase activity measured in buffer and 2 M NaCl
[c]$DZOase_{phys}$ = Diazoxonase activity measured under physiological conditions
[d]$CPOase_{phys}$ = Chlorpyrifos oxonase activity measured under physiological conditions
[e]$AREase_{LS}$ = Arylesterase activity measured in buffer
[f]From Richter et al. (2008)
[g]$DZOase_{HS}$ = Diazoxonase activity measured at 2 M NaCl, pH 8.5
(reproduced from Richter et al. 2009, with permission)

determining PON1 status and a new PON1 status protocol where rates of phenyl acetate hydrolysis at high salt are plotted against rates of 4-(chloromethyl)phenyl acetate (CMPA) in buffer alone (Richter et al. 2008, 2009). Both of these protocols resolve all three $PON1_{192}$ phenotypes; however the new assay with non-OP substrates is more suitable for laboratories that are not equipped for using highly toxic compounds.

Since many assay protocols have been used by different laboratories over the years, we determined conversion factors that will allow inter-conversion of rates of hydrolysis of one substrate to another for each of the three $PON1_{192}$ phenotypes (Richter et al. 2009) (Table 1). Also, we described how to determine physiologically relevant rates of in vivo hydrolysis of CPO and DZO. This new assay should encourage epidemiologists to measure the parameters that are important in relating genetic variability of PON1 to risk of disease or exposure.

We have shown previously that discrepancies between *PON1_{192}* SNP analysis and the functional PON1 status analysis can reveal mutations in the *PON1* gene that can be characterized by sequencing the entire *PON1* gene (Jarvik et al. 2003). The effects of polymorphisms in the 3'-untranslated region of the *PON1* gene have yet to be characterized. While there is still much to learn about the effects of *PON1* SNPs on expression, it will be important to couple such studies with the functional PON1 status analysis. Although the effects of some environmental influences on plasma PON1 levels are known (reviewed in Costa et al. 2005), the effects of epigenetic modifications on PON1 expression are yet to be explored.

Acknowledgments This work was supported by grants from the National Institute of Environmental Health Sciences ES09883, ES04696, ES07033, ES09601 – EPA: RD-83170901-OE and the National Heart, Lung, and Blood Institute, HL67406 and HL074366.

References

Brophy VH, Jarvik GP, Furlong CE (2002) PON1 Polymorphisms. In: Costa LG, Furlong CE (eds.). Paraoxonase (PON1) in Health and Disease: Basic and Clinical Aspects. Boston: Kluwer Academic Press. 53–77

Costa LG, Vitalone A, Cole TB and Furlong CE (2005) Modulation of paraoxonase (PON1) activity. Biochem Pharmacol 69:541–550

Chun CK, Ozer EA, Welsh MJ, Zabner J, Greenberg EP (2004) Inactivation of a Pseudomonas aeruginosa quorum sensing signal by human airway epithelia. Proc Natl Acad Sci USA 101:3587–3590

Cole TB, Walter BJ, Shih DM, Tward AD, Lusis AJ, Timchalk C, Richter RJ, Costa LG, Furlong CE (2005) Toxicity of chlorpyrifos and chlorpyrifos oxon in a transgenic mouse model of the human paraoxonase (PON1) Q192R polymorphism. Pharmacogenet Genomics 15:589–598

Costa LG, McDonald BE, Murphy SD, Omenn GS, Richter RJ, Motulsky AG, Furlong CE (1990) Serum paraoxonase and its influence on paraoxon and chlorpyrifos-oxon toxicity in rats. Toxicol Appl Pharmacol 103:66–76

Davies H, Richter RJ, Keifer M, Broomfield C, Sowalla J, Furlong CE (1996) The human serum paraoxonase polymorphism is reversed with diazoxon, soman and sarin. Nature Genet 14: 334–336

Deakin S, Leviev I, Brulhart-Meynet M-C, James RW (2003) Paraoxonase-1 promoter haplotypes and serum paraoxonase: a predominant role for polymorphic position −107, implicating the Sp1 transcription factor. Biochem J 372:643–649

Deakin SP, James RW (2004) Genetic and environmental factors modulating serum concentrations and activities of the antioxidant enzyme paraoxonase-1. Clin Sci (Lond.) 107:435–447

Eckerson HW, Wyte CM, La Du BN (1983) The human serum paraoxonase/arylesterase polymorphism. Am J Hum Genet 35:1126–1138

Furlong CE, Richter RJ, Li W-F, Brophy VH, Carlson C, Meider M, Nickerson D, Costa LG, Ranchalis J, Lusis AJ, Shih DM, Tward A, Jarvik GP (2008) The functional consequences of polymorphisms in the human PON1 gene. In: Mackness B, Mackness M, Aviram M, Paragh G (eds). The Paraoxonases: Their Role in Disease, Development and Xenobiotic Metabolism. Dordrecht, The Netherlands: Springer. 267–281

Furlong C, Holland N, Richter R, Bradman A, Ho A, Eskenazi B (2006) PON1 status of farmworker mothers and children as a predictor of organophosphate sensitivity. Pharmacogenet Genomics 16:183–190

James RW. (2006) A long and winding road: defining the biological role and clinical importance of paraoxonases. Clin Chem Lab Med 44:1052–1059

Jarvik GP, Rozek LS, Brophy VH, Hatsukami TS, Richter RJ, Schellenberg GD, Furlong CE (2000) Paraoxonase phenotype is a better predictor of vascular disease than PON1192 or PON155 genotype. Atheroscler Thromb Vasc Biol 20:2442–2447

Jarvik GP, Jampsa R, Richter RJ, Carlson C, Rieder M, Nickerson D, Furlong CE (2003) Novel paraoxonase (PON1) nonsense and missense mutations predicted by functional genomic assay of PON1 status. Pharmacogenetics 13:291–295

La Du BN (2003) Future studies of low-activity PON1 phenotype subjects may reveal how PON1 protects against cardiovascular disease, Arterioscler Thromb Vasc Biol 23:1317–1318

Lawlor DA, Day IN, Gaunt TR, Hinks LJ, Briggs PJ, Kiessling M, Timpson N, Smith GD, Ebrahim S (2004) The association of the PON1 Q192R polymorphism with coronary heart disease: findings from the British Women's Heart and Health cohort study and a meta-analysis. BMC Genet 5:17

Li W-F, Costa LG, Furlong CE (1993) Serum paraoxonase status: a major factor in determining resistance to organophosphates. J Toxicol Environ Health 40:337–346

Li W-F, Furlong CE, Costa LG (1995) Paraoxonase protects against chlorpyrifos toxicity in mice. Toxicol Lett 76:219–226

Li W-F, Costa LG, Richter RJ, Hagen T, Shih DM, Tward A, Lusis AJ, Furlong CE (2000) Catalytic efficiency determines the in vivo efficacy of PON1 for detoxifying organophosphates. Pharmacogenetics 10:767–780

Mackness B, Davies GK, Turkie W, Lee E, Roberts DH, Hill E, Roberts C, Durrington PN, Mackness MI (2001) Paraoxonase status in coronary heart disease: are activity and concentration more important than genotype? Arterioscler Thromb Vasc Biol 21:1451–1457

Ozer EA, Pezzulo A, Shih DM, Chun C, Furlong C, Lusis AJ, Greenberg EP, Zabner J (2005) Human and murine Paraoxonase 1 are host modulators of P. aeruginosa quorum-sensing. FEMS Microbiol Lett 253:29–37

Richter RJ, Furlong CE (1999) Determination of paraoxonase (PON1) status requires more than genotyping. Pharmacogenetics 9:745–753

Richter RJ, Jarvik GP, Furlong CE (2008) Determination of paraoxonase 1 status without the use of toxic organophosphate substrates. Circ Cardiovasc Genet 1:147–152 DOI: 10.1161/CIRCGENETICS.108.811638

Richter RJ, Jarvik GP, Furlong CE (2009) Paraoxonase 1 (PON1) status and substrate hydrolysis. Toxicol Appl Pharmacol 235:1–9

Shih DM, Gu L, Xia Y-R, Navab M, Li W-F, Hama S, Castellani LW, Furlong CE, Costa, LG, Fogelman AM, Lusis AJ (1998) Mice lacking serum paraoxonase are susceptible to organophosphate toxicity and atherosclerosis. Nature 394:284–287

Stoltz DA, Ozer EA, Taft PJ, Barry M, Liu L, Kiss PJ, Moninger TO, Parsek MR, Zabner J (2008) Drosophila are protected from Pseudomonas aeruginosa lethality by transgenic expression of paraoxonase-1. J Clin Invest 118:3123–3131 doi:10.1172/JCI35147

Wheeler JG, Keavney BD, Watkins H, Collins R, Danesh J (2004) Four paraoxonase gene polymorphisms in 11212 cases of coronary heart disease and 12786 controls: meta-analysis of 43 studies. Lancet 363:689–695.

Engineering Human PON1 in an *E. coli* Expression System

Stephanie M. Suzuki, Richard C. Stevens, Rebecca J. Richter,
Toby B. Cole, Sarah Park, Tamara C. Otto, Douglas M. Cerasoli,
David E. Lenz, and Clement E. Furlong

Abstract Expression and purification of recombinant human paraoxonase-1 (rHuPON1) from bacterial systems have proven elusive. Most systems for successful production of recombinant PON1 have relied on either eukaryotic expression in baculovirus or prokaryotic expression of synthetic, gene-shuffled rabbit–mouse–human PON1 hybrid molecules. We review here methods and protocols for the production of pure, native rHuPON1 using an *E. coli* expression system followed by conventional column chromatographic purification. The resulting rHuPON1 is stable, active, and capable of protecting *PON1* knockout mice (*PON1*$^{-/-}$) from exposure to high levels of the organophosphorus (OP) compound diazoxon. Bacterially-derived rHuPON1 can be produced in large quantities and lacks the glycosylation of eukaryotic systems that produces immunogenic complications when used as a therapeutic. The rHuPON1 should be useful for treating insecticide OP exposures and reducing risks of other diseases resulting from low PON1 status. The ease of mutagenesis in bacterial systems will also allow for the generation and screening of rHuPON1 variants with enhanced catalytic efficiencies against nerve agents and other OP compounds.

Keywords Recombinant PON1 · Nerve agents · Diazoxon · OP therapy · Engineered PON1 · *E. coli* expression · *PON1* knockout mice

1 Introduction

Each year there are approximately 2.5 million insecticide poisonings, resulting in 250,000 deaths (WHO, 1986, 1990). Many of these poisonings result from exposure to organophosphorus (OP) insecticides. In addition to accidental exposure to insecticides, the possibility exists of deliberate exposure to OP nerve agents or toxic industrial OPs. One approach for treating exposure to OP compounds has been the injection of stoichiometric scavengers such as butyrylcholinesterase (Ashani et al.

S.M. Suzuki (✉)
Departments of Medicine (Div. of Medical Genetics) and Genome Sciences, University of Washington, Seattle, Washington, USA
e-mail: stephis@u.washington.edu

S.T. Reddy (ed.), *Paraoxonases in Inflammation, Infection, and Toxicology*, Advances in Experimental Medicine and Biology 660, DOI 10.1007/978-1-60761-350-3_5,
© Humana Press, a part of Springer Science+Business Media, LLC 2010

1991; Broomfield et al. 1991). The advantage of using butyrylcholinesterase (BChE) as a therapeutic is that it will bind most toxic OP compounds; however, the disadvantage of using a stoichiometric scavenger is that it binds only a single molecule of an OP per large protein molecule. An attractive therapeutic alternative is the use of catalytic scavengers that hydrolyze many OP molecules per each injected protein molecule (Li et al. 2000; Lenz et al. 2007).

Paraoxonase-1 (PON1) is a leading candidate for use as a catalytic scavenger of toxic OP compounds. PON1 is capable of hydrolyzing a wide range of substrates including the OP pesticide products paraoxon (PO), diazoxon (DZO) and chlorpyrifos oxon (CPO), as well as nerve agents such as VX, VR, sarin and soman (Davies et al. 1996; Lenz et al. 2007). One requirement for a useful catalytic scavenger will be minimal immune reaction to the injected protein. The most promising approach to achieve this goal will be the use of engineered proteins of human origin. Other enzymes, such as bacterial phosphotriesterases, are capable of inactivating organophosphate toxins, but these are not of human origin (reviewed in Raushel, 2002). As PON1 is a native human protein, it will be much less likely to elicit an immune response compared to OP hydrolases from other organisms. OP hydrolysis is generally viewed as a promiscuous function of PON1 – its actual biological role is still debated (James, 2006), though it appears to be important for protecting against cardiovascular disease (Jarvik et al. 2000) and *Pseudomonas* infection (Ozer et al. 2005; Stoltz et al. 2008). PON1's antioxidant capability and association with HDL suggests that injectable, therapeutic rHuPON1 will have other possible applications such as prevention of blood vessel re-occlusion following intervention. The Q192R polymorphism in human PON1 affects the catalytic efficiency of hydrolysis for some OP substrates. Rabbit PON1 with lysine (K) at position 192 has a very high hydrolysis of chlorpyrifos oxon (Furlong et al. 1989; Hassett et al. 1991).

An effective catalytic scavenger must have a high catalytic efficiency in order to detoxify and protect against specific compounds (Li et al. 2000). For many years, it was thought that high levels of PON1 would protect against exposure to PO, from which PON1 derives its name. However, after the knockout mouse model was developed, this was shown to be incorrect. The $PON1^{-/-}$ mice were significantly more susceptive to CPO (Shih et al. 1998) and DZO (Li et al. 2000); however, they did not differ from wild-type mice in their susceptibility to PO. Similar results were found when mice were injected with purified human PON1 enzyme to test protection in vivo. The $PON^{-/-}$ mice were protected from inhibition of brain acetylcholinesterase when given PON_{R192} or PON_{Q192} and then dosed with either CPO or DZO. Neither purified PON alloform was able to protect mice against exposure to PO. The catalytic efficiency was determined for hydrolysis of CPO, DZO and PO for both PON_{R192} and PON_{Q192}. Both PON1s had nearly equivalent catalytic efficiency for DZO, reflected in the equal protection provided when injected. PON_{R192} had higher catalytic efficiency for hydrolysis of CPO, and provided better protection than PON_{Q192} against CPO exposure. The $PON1_{R192}$ alloform was nearly nine times more efficient than PON_{Q192} at hydrolyzing PO, but the overall catalytic efficiency was so low for both alloforms that neither could protect against PO exposure (Li et al. 2000). These experiments provided convincing evidence that engineered increases in catalytic efficiency will be required for protection against PO (and most likely nerve agents as well). We estimate that a ten-fold increase in catalytic efficiency of $PON1_{R192}$ will be sufficient to provide in vivo protection against PO exposure (Li et al. 2000).

Based on this previous work, the three important requirements for the development of a catalytic scavenger are minimal immunogenicity, a catalytic efficiency sufficient to protect against the specific compound and a good half-life following injection. With PON1, the development of rapid screening protocols for identifying variants with improved catalytic efficiency of hydrolysis is necessary. To prevent decoration of the scavenger with immunogenic carbohydrate chains, PON1 will be best expressed in a non-eukaryotic system. If active recombinant human PON1 (rHuPON1) could be expressed in an *E. coli* system, two of these goals would be realized. This system would also provide a more convenient system for scaled-up production and for rapid screening of variants with increased catalytic efficiencies.

The difficulty of using *E. coli* as a host has been noted previously (Brushia et al. 2001). In attempts to produce a more soluble version of PON1, Aharoni et al. (2004) used a gene-shuffling protocol to express a soluble variant of PON1 that was used to determine the first crystal structure of a PON1 enzyme (Harel et al. 2004). The resulting PON1 sequence was closer to rabbit PON1 than human PON1, with 91% identity to wild-type rabbit PON1. This hybrid protein would most likely not be ideal for use as a prophylactic or therapeutic drug in humans; to avoid immunological complications it would be desirable to have a protein therapeutic that is as close as possible to the native human sequence, with minimal changes necessary for increasing catalytic efficiency.

The purification of PON1, a highly hydrophobic protein, has also proved to be an obstacle in the development of a catalytic scavenger. Many of the early PON1 purification efforts used human serum as starting material, and produced quite pure PON1 following several chromatographic steps (Gan et al. 1991; Furlong et al. 1991). Recent improvements have reduced the number of steps required to purify human plasma PON1. Sinan et al. (2006) purified plasma PON1 with two steps, precipitation and hydrophobic interaction chromatography.

2 Expression and Purification of rHuPON1 from *E. coli*

To work out conditions for producing active human PON1 in an *E. coli* expression system, we expressed a GST-tagged rHuPON1 protein and demonstrated that active rHuPON1 fusion protein could be expressed and purified from *E. coli*. This expression system also allowed us to examine the effects of amino acid substitutions on the catalytic efficiency of the rHuPON1. The rabbit PON1 protein sequence (Hassett et al. 1991) suggested the possibility that substituting the 192 amino acid with lysine would lead to increased catalytic efficiency against OPs. We knew that a single amino acid substitution could produce a large change in catalytic rates, based on the variation seen in humans with the two natural $PON1_{192}$ alloforms. The first *E. coli* system used the popular pET vectors and produced active rHuPON1 with

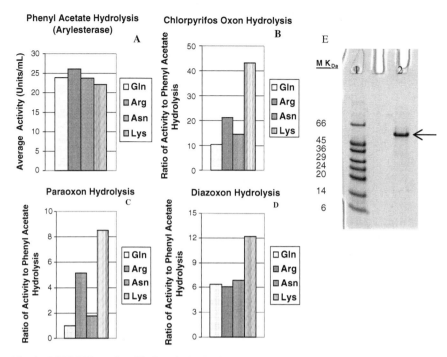

Fig. 1 GST-PON1 produced in *E. coli*. (**a**) Equivalent rates of arylesterase activity. (**b, c, d**) Ratios of rates of hydrolysis of chlorpyrifos oxon, paraoxon and diazoxon, respectively, to rates of phenyl acetate hydrolysis in rHuPON1 variants with Q, R, N or K at position 192. (**e**) SDS gel electrophoretic analysis of purified rHuPON1$_{Q192}$ fusion protein (*arrow*). Lane 1, molecular weight markers; Lane 2, purified rHuPON1$_{Q192}$ fusion protein

a GST tag for ease of purification. Both native human PON1$_{192}$ alloforms, as well as the K192 variant, were generated and characterized. Since the Q192 alloform of plasma PON1 is more catalytically active against nerve agents than the R192 alloform, a variant with asparagine (N) at position 192 was also expressed and characterized. Because the PON1$_{192}$ polymorphism has little effect on the activity of the enzyme towards phenyl acetate, we used the ratio of OP hydrolytic substrate activity to phenyl acetate hydrolytic activity for all four constructs. The PON1$_{K192}$ alloform did indeed show an increased activity against the three OPs tested (CPO, PO and DZO) relative to phenyl acetate (Fig. 1).

We have recently purified and characterized *E. coli*-generated native, wild-type human rHuPON1$_{Q192}$ and rHuPON1$_{R192}$ alloforms, as well as the rHuPON1$_{K192}$ variant. These rHuPON1$_{192}$ proteins were purified to greater than 90% homogeneity (Stevens et al. 2008). A new expression system was developed to produce the untagged native rHuPON1 which would be ideal for a therapeutic protein. We used the Eurogentec Staby system, which employs a toxin–antitoxin method to force the expressing cells to retain the plasmid (Bernard and Couturier, 1992). Another plasmid, pRARE, that encodes tRNAs common in mammalian proteins but rare in bacterial proteins, was co-transformed to increase yields (Invitrogen). This *E. coli* expression system produced active protein. Cells were grown in a 14 L fermenter, harvested, then lysed with glass beads in a detergent-containing buffer and purified using column chromatography. The final series of columns utilized multiple ion exchange DEAE columns with and without detergent at different pH values and a hydrophobic interaction column with a detergent-containing buffer elution (Fig. 2).

3 Kinetic Properties of rHuPON1$_{192}$ Variants

Kinetic analysis was performed using all three variants of PON1. Notably, PON1$_{K192}$ had a catalytic activity approximately twice that of PON1$_{R192}$ for several OP substrates (CPO, DZO, PO). However, the increased efficiency towards PO is most likely not yet high enough to protect against a PO exposure in the *PON1$^{-/-}$* mouse model. The K_m values for PON1$_{K192}$ were higher relative to PON1$_{R192}$; however, the V_{max} was also much higher for the PON1$_{K192}$ variant. Catalytic efficiencies for the two natural rHuPON1 human alloforms were similar to those of PON1 purified from serum (Li et al. 2000) (Table 1). Initial turnover values for the rHuPON1 variants were carried out with VX and VR nerve agents (Table 2). None of the rHuPON1 variants could be saturated at 1.4 mM VX and only rHuPON1$_{Q192}$ could be saturated at 1.4 mM VR. The K_m of rHuPON1$_{Q192}$ for VR hydrolysis was 0.72 mM and the K_{cat} was 29.8 min^{-1}.

4 Testing rHuPON1K$_{192}$ as a Therapeutic

To test the in vivo efficacy of using rHuPON1 as a catalytic scavenger, experiments were carried out using *PON1$^{-/-}$* knockout mice. These mice were chosen for tests of the rHuPON1 since there is no DZOase activity in these mice. Mice were housed in either barrier or modified SPF (specific pathogen free) facilities with 12 h dark–light cycles and free access to food and water. All experiments were carried out in accordance with the National Research Council Guide for the Care and Use of Laboratory Animals, as adopted by the National Institutes of Health. Animal use protocols were approved by the Institutional Animal Care and Use Committee at the University of Washington.

To determine if the *E. coli*-produced rHuPON1 was nontoxic, 1.12 DZOase units of PON1$_{K192}$ were injected intraperitoneally (i.p.) into each of three of the *PON1$^{-/-}$* mice. The mice showed no signs of physical symptoms. The increase in serum PON1 DZOase levels was about half that of wild-type levels. This increase in levels peaked about 8 h after injection. Antibodies to PON1 could not be detected in mouse serum by ELISA four months after injection. Mice were still alive with no symptoms 12 months following injection.

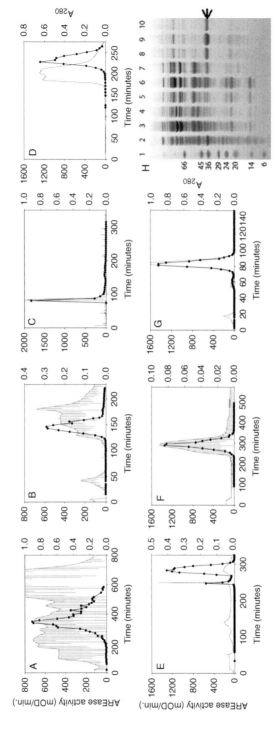

Fig. 2 Purification of the untagged rHuPON1$_{K192}$ variant. Column steps were: (**a**) DEAE 1; (**b**) DEAE 2; (**c**) hydroxyapatite (HA); (**d**) DEAE 3; (**e**) hydrophobic interaction column (HIC); (**f**) gel filtration (GF); and (**g**) DEAE 4. The activity traces are shown as *black lines* with data points and the A280 traces are shown as continuous *grey lines* with fraction ticks. (**h**) SDS-PAGE analysis of the pooled fractions from each step: Lane 1, molecular weight markers; Lane 2, cell extract; Lane 3, DEAE 1; Lane 4, desalting column; Lane 5, DEAE 2; Lane 6, HA; Lane 7, DEAE 3; Lane 8, HIC; Lane 9, GF and Lane 10, DEAE 4. *Arrow* represents the mobility of rHuPON1. Figure reproduced from Stevens et al. 2008

Table 1 Kinetic analysis of substrate hydrolysis for insecticide OPs

Substrate	PON1 variant[a]	K_m (mM)	V_{max} (U/mg)	V_{max}/K_m
Paraoxon	rHuPON1$_{K192}$	0.925 ± 0.029	11.66 ± 0.06	12.61 ± 0.32
	rHuPON1$_{R192}$	0.868 ± 0.016	6.61 ± 0.18	7.62 ± 0.23
	plasma HuPON1$_{R192}$	0.52	3.26	6.27
Diazoxon	rHuPON1$_{K192}$	2.57 ± 0.62	301 ± 51	118 ± 10
	rHuPON1$_{R192}$	1.33 ± 0.08	119 ± 5	89.8 ± 2.3
	plasma HuPON1$_{R192}$	1.02	79	77
Chlorpyrifos oxon	rHuPON1$_{K192}$	0.317 ± 0.02	245 ± 3	777 ± 66
	rHuPON1$_{R192}$	0.131 ± 0.005	34.7 ± 0.3	266 ± 12
	plasma HuPON1$_{R192}$	0.25	64	256
Phenyl acetate	rHuPON1$_{K192}$	3.22 ± 0.79	3020 ± 300	966 ± 166
	rHuPON1$_{R192}$	0.957 ± 0.02	680 ± 0	711 ± 16

[a] Data for plasma PON1 are from Li et al. 2000; data for rHuPON1 are from Stevens et al. 2008.

Table 2 Turnover numbers for substrate hydrolysis of nerve agent OPs

Substrate	PON1 variant	Turnover (min^{-1})
VR	rHuPON1$_{K192}$	5.2
	rHuPON1$_{Q192}$	19.4
	rHuPON1$_{R192}$	6.8
VX	rHuPON1$_{K192}$	5.74
	rHuPON1$_{Q192}$	33.4
	rHuPON1$_{R192}$	10.6

The next question was to determine if injected rHuPON1 could protect *PON1*$^{-/-}$ mice from OP exposure. Administration of 192 μg of rHuPON1$_{K192}$ (3.91 units of total DZOase activity) i.p. and intramuscularly (i.m.) to two mice was followed by dermal exposure to 1 mg/kg DZO 48 h following injection. Six hours following exposure, the brain cholinesterase levels of the mice were determined. Mice that were injected with the rHuPON1$_{K192}$ had no observable inhibition of ChE while mice exposed to DZO and not receiving rHuPON1$_{K192}$ had an approximately 50% reduction in brain ChE. The next question to be addressed was whether rHuPON1 could be injected 10 min after dermal exposure to high levels of DZO (i.e. >2 × LD$_{50}$). One mouse was exposed to 3 mg/kg and the other to 7 mg/kg. However, the mice not only survived, but showed fewer cholinergic symptoms than control mice that were exposed to a lower dose (1.5 mg/kg).

5 Conclusions

We have previously proposed the use of PON1 as a catalytic scavenger for treating OP exposures based on experiments where purified plasma PON1 provided protection against OP exposure in wild-type mice (Li et al. 2000). This series of experiments has provided evidence of the suitability of using rHuPON1 for therapeutic or prophylactic protection against OP exposure. These experiments have demonstrated that native human PON1 can be produced in bacterial cells and processed to sufficient purity for injection without side effects. This *E. coli* expression system will allow for high throughput screening for mutants with improved catalytic efficiency and for scaled-up production of therapeutic PON1.

Acknowledgments This work was supported by National Institutes of Health Grants ES09883, ES04696, ES07033, and ES09601/EPA: RD-83170901, and a grant from the University of Washington Center for Process Analytical Chemistry (CFO1). This work was supported in part by the Defense Threat Reduction Agency (DEL, DMC and TCO). Opinions, interpretations, conclusions, and recommendations are those of the authors and are not necessarily endorsed by the US Army.

References

Aharoni A, Gaidukov L, Yagur S, Toker L, Silman I, Tawfik DS (2004) Directed evolution of mammalian paraoxonases PON1 and PON3 for bacterial expression and catalytic specialization. Proc Natl Acad Sci USA 101:482–487

Ashani Y, Shapira S, Levy D, Wolfe AD, Doctor BP, Raveh L (1991) Butyrylcholinesterase and acetylcholinesterase prophylaxis against soman poisoning in mice. Biochem Pharmacol 41: 37–41

Bernard P, Couturier M (1992) Cell killing by the F plasmid CcdB protein involves poisoning of DNA-topoisomerase II complexes. J Mol Biol 226:735–745

Broomfield CA, Maxwell DM, Solana RP, Castro CA, Finger AV, Lenz DE (1991) Protection by butyrylcholinesterase against organophosphorus poisoning in nonhuman primates. J Pharmacol Exp Ther 259(2):633–638

Brushia RJ, Forte TM, Oda MN, La Du BN, Bielicki JK (2001) Baculovirus-mediated expression and purification of human serum paraoxonase 1A. J Lipid Res 42:951–958.

Davies HG, Richter RJ, Keifer M, Broomfield CA, Sowalla J, Furlong CE (1996) The effect of the human serum paraoxonase polymorphism is reversed with diazoxon, soman and sarin. Nat Genet 14:334–336

Furlong CE, Richter RJ, Seidel SL, Costa LG, Motulsky AG (1989) Spectrophotometric assays for the enzymatic hydrolysis of the active metabolites of chlorpyrifos and parathion by plasma paraoxonase/arylesterase. Anal Biochem 180:242–247

Furlong CE, Richter RJ, Chapline C, Crabb JW (1991) Purification of rabbit and human serum paraoxonase. Biochemistry 30:10133–10140

Gan KN, Smolen A, Eckerson HW, La Du BN (1991) Purification of human serum paraoxonase/arylesterase: Evidence for one esterase catalyzing both activities. Drug Metabolism and Dispos 19:100–106

Harel M, Aharoni A, Gaidukov L, Brumshtein B, Khersonsky O, Meged R, Dvir H, Ravelli RBG, McCarthy A, Toker L, Silman I, Sussman JL, Tawfik DS (2004) Structure and evolution of the serum paraoxonase family of detoxifying and anti-atherosclerotic enzymes. Nat Struct Mol Biol 11:412–419

Hassett C, Richter RJ, Humbert R, Chapline C, Crabb JW, Omiecinski CJ, Furlong CE (1991) Characterization of cDNA clones encoding rabbit and human serum paraoxonase: The mature protein retains its signal sequence. Biochemistry 30:10141–10149

James RW. (2006) A long and winding road: defining the biological role and clinical importance of paraoxonases. Clin Chem Lab Med 44:1052–1059

Jarvik GP, Rozek LS, Brophy VH, Hatsukami TS, Richter RJ, Schellenberg GD, Furlong CE (2000) Paraoxonase (PON1) phenotype is a better predictor of vascular disease than is PON1(192) or PON1(55) genotype. Arterioscler Thromb Vasc Biol 20(11):2441–2447

Lenz DE, Yeung D, Smith JR, Sweeney RE, Lumley LA, Cerasoli D.(2007) Stoichiometric and catalytic scavengers as protection against nerve agent toxicity: a mini review. Toxicology 233:31–39

Li WF, Costa LG, Richter RJ, Hagen T, Shih DM, Tward A, Lusis AJ, Furlong CE (2000) Catalytic efficiency determines the in-vivo efficacy of PON1 for detoxifying organophosphorus compounds. Pharmacogenetics 10:767–779

Ozer EA, Pezzulo A, Shih DM, Chun C, Furlong C, Lusis AJ, Greenberg EP, Zabner J (2005) Human and murine paraoxonase 1 are host modulators of Pseudomonas aeruginosa quorum-sensing. Fems Microbiol Lett 253:29–37

Raushel FM (2002) Bacterial detoxification of organophosphate nerve agents. Curr Opin Microbiol 5(3):288–295

Shih DM, Gu LJ, Xia YR, Navab M, Li WF, Hama S, Castellani LW, Furlong CE, Costa LG, Fogelman AM, Lusis AJ (1998) Mice lacking serum paraoxonase are susceptible to organophosphate toxicity and atherosclerosis. Nature 394:284–287

Sinan S, Kockar F, Arslan O (2006) Novel purification strategy for human PON1 and inhibition of the activity by cephalosporin and aminoglikozide derived antibiotics. Biochimie 88:565–574

Stevens RC, Suzuki SM, Cole TB, Park SS, Richter RJ, Furlong CE (2008) Engineered recombinant human paraoxonase 1 (rHuPON1) purified from *Escherichia coli* protects against organophosphate poisoning. Proc Natl Acad Sci USA 105(35):12780–12784

Stoltz DA, Ozer EA, Taft PJ, Barry M, Liu L, Kiss PJ, Moninger TO, Parsek MR, Zabner J (2008) Drosophila are protected from Pseudomonas aeruginosa lethality by transgenic expression of paraoxonase-1. J Clin Invest 118:3123–3131

World Health Organization. Informal consultation on planning strategy for the prevention of pesticide poisoning. Geneva, 25-29 November 1985. WHO/VBC/86.926. (Geneva: WHO, 1986)

World Health Organization. Public health impact of pesticides used in agriculture (Geneva: WHO, 1990)

The Toxicity of Mixtures of Specific Organophosphate Compounds is Modulated by Paraoxonase 1 Status

Toby B. Cole, Karen Jansen, Sarah Park, Wan-Fen Li, Clement E. Furlong, and Lucio G. Costa

Abstract Most chemical exposures involve complex mixtures. The role of paraoxonase 1 (PON1) and the Q192R polymorphism in the detoxication of individual organophosphorous (OP) compounds has been well-established. The extent to which PON1 protects against a given OP is determined by its catalytic efficiency. We used a humanized transgenic mouse model of the Q192R polymorphism to demonstrate that PON1 modulates the toxicity of OP mixtures by altering the activity of another detoxication enzyme, carboxylesterase (CaE). Chlorpyrifos oxon (CPO), diazoxon (DZO), and paraoxon (PO) are potent inhibitors of CaE, both in vitro and in vivo. We hypothesized that exposure of mice to these OPs would increase their sensitivity to the CaE substrate, malaoxon (MO), and that the degree of effect would vary among PON1 genotypes if the OP was a physiologically relevant PON1 substrate. When wild-type mice were exposed dermally to CPO, DZO, or PO and then, after 4 h, to different doses of MO, the toxicity of MO was increased compared to mice that received MO alone. The potentiation of MO toxicity by CPO and DZO was higher in PON1 knockout mice, which are less able to detoxify CPO or DZO. Potentiation by CPO was higher in Q192 mice than in R192 mice due to the decreased ability of $PON1_{Q192}$ to detoxify CPO. Potentiation by DZO was similar in the Q192 and R192 mice, due to their equivalent effectiveness at detoxifying DZO. PO exposure resulted in equivalent potentiation of MO toxicity among all four genotypes. These results indicate that PON1 status modulates the ability of CaE to detoxicate OP compounds from specific mixed insecticide exposures. PON1 status can also impact the capacity to metabolize drugs or other CaE substrates following insecticide exposure.

Keywords Mixed exposures · Chlorpyrifos · Chlorpyrifos oxon · Diazinon · Diazoxon · Malathion · Malaoxon · Parathion · Paraoxon · Pyrethroids · Tricresyl phosphate · Carboxylesterase · Paraoxonase 1 (PON1)

C.E. Furlong (✉)
Department of Medicine and Genome Sciences, University of Washington, Seattle, WA, USA
e-mail: clem@u.washington.edu

S.T. Reddy (ed.), *Paraoxonases in Inflammation, Infection, and Toxicology*, Advances in Experimental Medicine and Biology 660, DOI 10.1007/978-1-60761-350-3_6,

1 PON1 and Detoxication of OP Compounds

The involvement of paraoxonase 1 (PON1) in the detoxication of organophospho-rus (OP) compounds has been well-documented (reviewed in Costa, 2006; Furlong et al., 2008). *PON1* knockout (*PON1$^{-/-}$*) mice are dramatically more sensitive than wild-type (*PON1$^{+/+}$*) mice to the toxicity of chlorpyrifos oxon (CPO) and diazoxon (DZO) and to a lesser extent the parent phosphorothioates, chlorpyrifos and diazi-non (Shih et al. 1998; Li et al. 2000; Cole et al. 2005). Injection of purified plasma PON1 protein (Main, 1956; Costa et al. 1990; Li et al. 1995, 2000) or, more recently, recombinant PON1 (Stevens et al. 2008), increased the resistance of rats and/or mice to OP toxicity. The extent of protection was shown to be dependent on the catalytic efficiency of PON1 hydrolysis for the respective OP compounds (Li et al. 2000). The Q192R amino acid polymorphism of PON1 (hPON1$_{Q192R}$) affects the catalytic efficiency of hydrolysis for some substrates, but not others (Hassett et al. 1991; Adkins et al. 1993; Davies et al. 1996; Li et al. 2000). For DZO, the hPON1$_{Q192}$ and hPON1$_{R192}$ alloforms had equivalent catalytic efficiencies measured in vitro and injection of the hPON1$_{Q192}$ and hPON1$_{R192}$ alloforms provided equivalent pro-tection in vivo (Li et al. 2000). For chlorpyrifos CPO, the hPON1$_{R192}$ alloform had a higher catalytic efficiency of hydrolysis than the hPON1$_{Q192}$ alloform in vitro and also provided better protection than hPON1$_{Q192}$ in vivo (Li et al. 2000). For paraoxon (PO), the catalytic efficiency measured in vitro was very low, and PON1 did not provide any protection in vivo (Li et al. 2000). Further evidence came from a study of humanized PON1 transgenic mice, in which the endogenous mouse *PON1* gene was removed and human *PON1* transgenes were inserted that encoded either hPON1$_{Q192}$ or hPON1$_{R192}$ (Cole et al. 2003). The hPON1$_{Q192}$ mice were much more sensitive than hPON1$_{R192}$ mice to CPO and, to a lesser extent, chlorpyrifos (Cole et al. 2005).

2 PON1 Status

In addition to the hPON1$_{Q192R}$ amino acid polymorphism, activity levels of plasma PON1 vary tremendously, as much as 15-fold among individuals of the same hPON1$_{Q192R}$ genotype (Furlong et al. 2006; Furlong, 2007). PON1 levels are also very low and variable in newborns, and reach adult levels between 6 months and 2 years of age (Cole et al. 2003; Eskenazi et al. 2008). "PON1 status" is a term that was introduced to take into account both the hPON1$_{Q192R}$ polymorphism and the level of plasma PON1 activity (Li et al. 1993; Richter and Furlong, 1999). PON1 status has been determined primarily through the use of a two-substrate assay that compares plasma rates of DZO hydrolysis in the presence of high salt to plasma rates of PO hydrolysis (Li et al. 1993; Richter and Furlong, 1999; Costa et al. 1999; Jarvik et al. 2003; Huen et al. 2009). More recently, Richter et al. (2008, 2009) developed a pro-tocol that uses the non-toxic substrates phenyl acetate and 4-(chloromethyl)phenyl acetate to determine an individual's PON1 status.

3 Toxicity of OP Mixtures

Chemical exposures are likely to involve multiple different types of compounds and different routes (e.g., oral and dermal). The assumption of the EPA cumulative risk assessment for the OPs as a class of compounds was dose additivity, with the rationale that the OPs share a common mechanism of action (US EPA, 1999, 2002, 2006). However, numerous studies have reported greater-than-additive effects of combinations of OP compounds. Of particular relevance are early studies demonstrating that toxicity associated with exposure to malathion or its oxon metabolite, malaoxon (MO), was potentiated when the exposure occurred in combination with compounds that inhibit carboxylesterases (CaEs) (Aldridge, 1954; Cook et al. 1957; Dubois, 1958; Murphy et al. 1959; Seume and O'Brien, 1960; Casida et al. 1961, 1963; Cohen and Murphy, 1971a;b; Verschoyle et al. 1982). Malathion is converted in the liver to MO, which can be a potent inhibitor of AChE (DuBois et al. 1953; March et al. 1956; Murphy and DuBois, 1957; O'Brien, 1957), yet potentiation of malathion/MO toxicity was observed even at doses that would not normally inhibit acetylcholinesterase activity (Dubois, 1969; Su et al. 1971). CaEs hydrolyze the carboxylic esters of malathion and MO (March et al. 1956; O'Brien, 1957; Cook and Yip, 1958; Chen et al. 1969). In vitro, MO can undergo hydrolysis by CaEs, but can also bind to CaEs resulting in irreversible inhibition (Main and Dauterman, 1967). Other OP compounds, most notably CPO, DZO, and PO, are not hydrolyzed by CaEs, but instead bind to CaEs and other serine esterases (B-esterases) stoichiometrically and irreversibly, allowing the CaEs to act as stoichiometric scavengers of OP compounds and inhibiting the CaEs in the process (Su et al. 1971; Ramakrishna and Ramachandran, 1978; Chambers et al. 1990; Buratti and Testai, 2005). Tang and Chambers (1999) also found that triorthocresyl phosphate (TOCP) pretreatment potentiated PO toxicity, supporting the role of CaE in the detoxication of PO. CaE activity is highest in the liver, gastrointestinal tract, and brain, with interindividual variability as high as 44-fold among samples of human liver microsomes (Hosokawa et al. 1995; Satoh and Hosokawa, 2006). Rodents, but not humans, possess significant plasma CaE activity (Williams et al. 1989; Li et al. 2005).

Several more recent studies examined the toxicity of OP mixtures. Moser et al. (2005, 2006), using concurrent exposure to a mixture of five OP compounds, found greater-than-additive effects on the potentiation of malathion toxicity, using AChE inhibition and behavioral changes as endpoints. Timchalk et al. (2005) used a binary mixture of diazinon and chlorpyrifos in the rat, and reported additive effects on AChE at low doses (15 mg/kg), and interactive effects at a higher dose (60 mg/kg).

4 Effect of PON1 on the Interactive Toxicity of OP Mixtures

We performed a series of experiments to demonstrate that differences in OP detoxication between the $hPON1_{Q192}$ and $hPON1_{R192}$ alloforms can affect the interactive toxicity of chemical mixtures (Jansen et al. 2009). As reported below, the OP compounds CPO, DZO, and PO bind to CaE and inhibit its activity. By virtue of their differential detoxication of CPO, DZO, and PO, $hPON1_{Q192}$ and $hPON1_{R192}$ modulate the degree of this OP-mediated CaE inhibition. As a result, in a combined or sequential exposure PON1 status can modulate the interactive toxicity of OP compounds, even when one of the compounds is not metabolized directly by PON1. We demonstrated this to be the case for the toxicity of MO, which is not a physiologically-relevant PON1 substrate, when combined with exposure to DZO and CPO, which are physiologically-relevant PON1 substrates.

5 Inhibition of CaE by OP Compounds In Vitro

Inhibition of CaE and AChE by OP compounds was measured in liver and brain homogenates and plasma prepared from wild-type ($PON1^{+/+}$; B6.129) mice. Liver, plasma, and brain samples were incubated with CPO, DZO, PO, or MO for 30 minutes at 23°C, followed by measurement of CaE or AChE activity. CPO, DZO, and PO were relatively potent inhibitors of CaE and AChE, with IC_{50} values in the low nM range (Table 1).

Table 1 In vitro IC_{50} values of CPO, DZO, PO, and MO for plasma and liver CaE and brain AChE

Tissue	Organophosphorus insecticide (nM)[a]			
	CPO	DZO	PO	MO
Plasma (CaE)	33.7 ± 8.3	16.6 ± 4.2	18.7 ± 2.1	5976.0 ± 495
Liver (CaE)	4.3 ± 0.8	3.9 ± 0.6	1.1 ± 0.3	308.8 ± 29.2
Brain (AChE)	14.0 ± 1.9	96.9 ± 7.6	9.5 ± 1.3	74.6 ± 7.8

[a] IC_{50} values (mean $\pm$ SEM). Data from Jansen et al. (2009) with permission

6 Inhibition of CaE by OP Compounds In Vivo

Transgenic and knockout mice were used to address whether OP compounds inhibited CaE in vivo and whether PON1 is involved in modulating the toxicity of mixtures of OP compounds. $PON1^{-/-}$ mice (Shih et al. 1998) and mice expressing either the human $hPON1_{R192}$ or $hPON1_{Q192}$ transgene in place of endogenous mouse PON1 (Cole et al. 2003, 2005) were provided by Drs. Diana M. Shih, Aaron Tward and Aldons J. Lusis (UCLA, Los Angeles, CA). $PON1^{+/+}$ mice were bred from the same congenic B6.129 strain background. Mice were housed in SPF (specific pathogen-free) facilities with a 12-h dark–light cycle and unlimited access to food and water. Experiments were carried out in accordance with the National Research Council Guide for the Care and Use of Laboratory Animals, as adopted by the National Institutes of Health, and were approved by the Institutional Animal Care and Use Committee at the University of Washington.

To determine plasma PON1 levels in the mice, saphenous-vein plasma was used to measure the rates of hydrolysis of the alloform-neutral substrates, phenyl acetate (Furlong et al. 1989, 1993, 2006) or DZO, which is alloform-neutral at physiological salt concentrations (Richter and Furlong, 1999). Arylesterase (AREase) and diazoxonase (DZOase) assays were carried out in a microtiter plate reader (SpectraMax Plus, Molecular Devices), and the initial linear rates of hydrolysis were used to calculate units of activity per ml plasma. As seen previously (Cole et al. 2003, 2005), plasma from the $PON1^{-/-}$ mice had some background AREase activity that was not due to PON1, whereas DZOase activity was essentially absent from $PON1^{-/-}$ mouse plasma (Table 2). Plasma PON1 levels were about 50% higher in $hPON1_{Q192}$ mice compared to $hPON1_{R192}$ mice (Table 2; Jansen et al. 2009).

Table 2 Serum PON1 levels in the experimental mice [a]

	$hPON1_{Q192}$	$hPON1_{R192}$	$PON1^{+/+}$	$PON1^{-/-}$
AREase (Units/ml)	76.49 ± 3.65	53.07 ± 3.50	65.42 ± 3.00	13.60 ± 1.01
DZOase (Units/ml)	10.95 ± 0.60	4.93 ± 0.19	6.42 ± 0.61	0.06 ± 0.02

[a] Data from Jansen et al. (2009) with permission

To determine the time course of CaE inhibition by OP compounds in vivo, mice were exposed dermally (1 µl/g body weight) to 0.5 mg/kg DZO or 0.75 mg/kg CPO, or to 0.35 mg/kg PO, which inhibits CaE but is not a physiologically-relevant PON1 substrate. Plasma CaE inhibition was maximal 4 hours after exposure (Fig. 1; Jansen et al. 2009). For CPO exposure, the order of sensitivity to plasma CaE inhibition at 4 hours (from greatest to least inhibition) was $PON1^{-/-} > hPON1_{Q192} > hPON1_{R192} > PON1^{+/+}$ (Fig. 1a), as expected based on their different catalytic efficiencies of CPO hydrolysis (Li et al. 2000). For DZO exposure, $PON1^{-/-}$ mice were more sensitive than $PON1^{+/+}$ mice to CaE inhibition, and $hPON1_{Q192}$ and $hPON1_{R192}$ mice had similar sensitivities to CaE inhibition (Fig. 1b) as expected based on their equivalent catalytic efficiencies of hydrolysis (Li et al. 2000). For exposure to PO, which is not a physiologically-relevant PON1 substrate, there were no differences in CaE inhibition among genotypes (Fig. 1c). At these concentrations of OPs, there was minimal to no inhibition of liver CaE (Jansen et al. 2009). Exposure to higher doses of OPs (0.50–0.75 mg/kg PO; 1.5–2.0 mg/kg DZO; 1.5–3.0 mg/kg CPO) resulted in inhibition of liver CaE and even more substantial inhibition of serum CaE (Jansen et al. 2009).

7 Effect of PON1 Status on the Toxicity of OP Compound Mixtures

The effects of CPO, DZO, and PO on subsequent toxicity of MO were determined by exposing mice dermally to 0.75 mg/kg CPO, 0.5 mg/kg DZO, or 0.35 mg/kg PO, followed 4 hours later (at the time of maximal CaE inhibition) by dermal exposure to

Fig. 1 Time course of
plasma CaE inhibition in
vivo, following exposure to
CPO, DZO, and PO. Time
course of plasma
carboxylesterase (CaE)
inhibition in $PON1^{+/+}$,
$PON1^{-/-}$, $hPON1_{Q192}$, and
$hPON1_{R192}$ mice (genotypes
as indicated) following
dermal exposure to
0.75 mg/kg CPO (**a**),
0.5 mg/kg DZO (**b**), or
0.35 mg/kg PO (**c**). Maximal
inhibition of CaE was at 4
hours. Results represent the
mean ± SEM ($n = 5$–10).
Reproduced from Jansen
et al. (2009) with permission

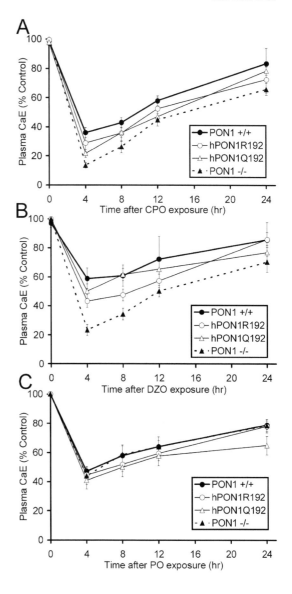

MO (Jansen et al. 2009). Pre-exposure to CPO, DZO, or PO inhibited plasma CaE,
and was associated with a significant ($p < 0.01$; multifactorial ANOVA) increase in
MO-mediated inhibition of brain and diaphragm AChE (Figs. 2, 3, and 4, compare a,
b vs. c, d and e, f). To assess whether the presence of PON1 affected this potentiation
of MO toxicity, $PON1^{+/+}$ mice were compared to $PON1^{-/-}$ mice for their sensitivity
to mixed OP exposures (Figs. 2c, d, 3c, d, and 4c, d). With pre-exposure to CPO
(Fig. 2c, d) or DZO (Fig. 3c, d), but not PO (Fig. 4c, d), AChE inhibition by MO was

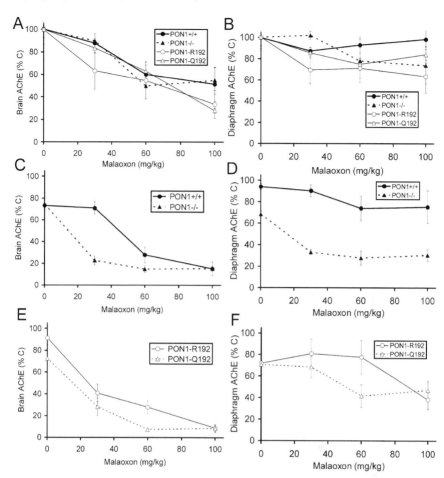

Fig. 2 Effect of CPO exposure (0.75 mg/kg) on subsequent toxicity of malaoxon (MO). Mice (genotypes as indicated) were exposed dermally to MO alone (**a, b**), or to CPO followed 4 hours later by MO exposure (**c, d, e, f**). AChE was measured in the brain (**a, c, e**) and diaphragm (**b, d, f**) 4 hours following the MO exposure. Results represent the mean ± SEM ($n = 4$). Reproduced from Jansen et al. (2009) with permission

significantly greater in $PON1^{-/-}$ mice than in $PON1^{+/+}$ mice ($p < 0.02$, CPO/MO; $p < 0.0001$, DZO/MO; $p = 0.87$, PO/MO), consistent with the known roles of PON1 in detoxication of CPO and DZO, but not PO (Li et al. 2000).

To assess whether the hPON1$_{Q192R}$ polymorphism affected the potentiation of MO toxicity, $hPON1_{Q192}$ and $hPON1_{R192}$ transgenic mice were compared for sensitivity to mixed OP exposures (Figs. 2e, f, 3e, f, and 4e, f). The results were consistent with the different catalytic efficiencies of hydrolysis of hPON1$_{Q192}$ and hPON1$_{R192}$ for CPO, DZO, and PO. Specifically, with pre-exposure to PO, AChE inhibition by MO was not affected by either the presence of PON1 (Fig. 4c, d) or

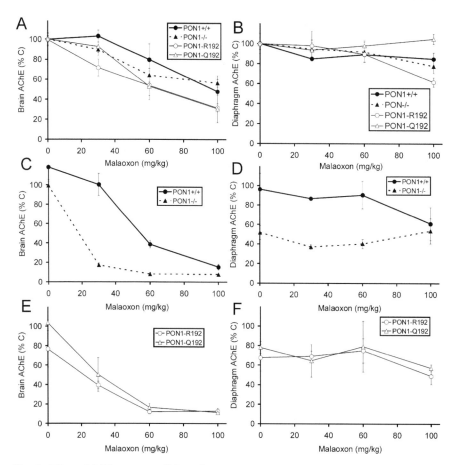

Fig. 3 Effect of DZO exposure (0.5 mg/kg) on subsequent toxicity of malaoxon (MO). Mice (genotypes as indicated) were exposed dermally to MO alone (**a**, **b**), or to DZO followed 4 hours later by MO exposure (**c**, **d**, **e**, **f**). AChE was measured in the brain (**a**, **c**, **e**) and diaphragm (**b**, **d**, **f**) 4 hours following the MO exposure. Results represent the mean ± SEM ($n = 4$). Reproduced from Jansen et al. (2009) with permission

by $hPON1_{Q192R}$genotype (Fig. 4e, f). With pre-exposure to DZO, AChE inhibition by MO was affected by the presence of PON1 (Fig. 3c, d), but there was no difference ($p = 0.13$) in modulation between the hPON1$_{Q192}$ and hPON1$_{R192}$ alloforms (Fig. 3e, f). In contrast, with pre-exposure to CPO, AChE inhibition by MO was affected by both the presence of PON1 (Fig. 2c, d) and by the hPON1$_{Q192R}$ polymorphism (Fig. 2e, f). With CPO pre-exposure, $hPON1_{Q192}$ mice had significantly ($p < 0.02$) greater inhibition of AChE by MO than did the $hPON1_{R192}$ mice (Fig. 2e, f). These results are consistent with a higher catalytic efficiency of CPO hydrolysis by the hPON1$_{R192}$ alloform compared to the hPON1$_{Q192}$ alloform, and with their equivalent catalytic efficiencies of DZO hydrolysis (Li et al. 2000).

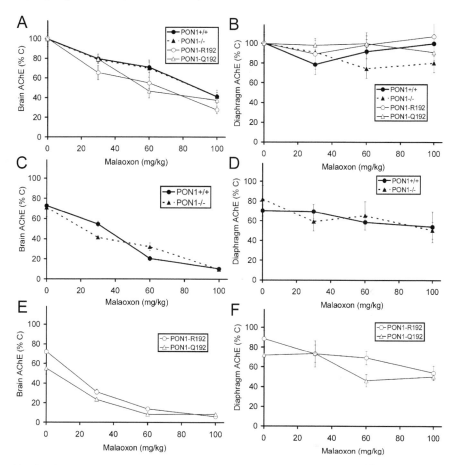

Fig. 4 Effect of PO exposure (0.35 mg/kg) on subsequent toxicity of malaoxon (MO). Mice (geno-types as indicated) were exposed dermally to MO alone (**a, b**), or to CPO followed 4 hours later by MO exposure (**c, d, e, f**). AChE was measured in the brain (**a, c, e**) and diaphragm (**b, d, f**) 4 hours following the MO exposure. Results represent the mean ± SEM ($n = 4$). Reproduced from Jansen et al. (2009) with permission

8 Conclusions

Clearly, PON1 status modulates the interactive toxicity of OP compounds. We demonstrated that CPO, DZO, and PO inhibit CaE in vitro and in vivo and increase MO toxicity in vivo, and that PON1 status modulates the degree of MO potentiation by virtue of its impact on the metabolism of CPO and DZO. The degree to which CPO, DZO, or PO inhibited CaE was predictive of their degree of potentiation of MO toxicity. Whereas PON1 had no affect on the potentiation of MO toxicity by PO, the absence of PON1 significantly increased the potentiation of MO toxicity by both CPO and DZO. These data indicate that interindividual differences in plasma PON1 levels would be important for determining sensitivity to mixed exposures involving diazinon/DZO and pesticides detoxified by the CaEs. Plasma PON1 levels are highly variable not just among individuals, but during development as well (Cole et al. 2003; Furlong et al. 2006). For mixed exposures involving chlorpyrifos/CPO, both plasma PON1 levels and $hPON1_{Q192R}$ genotype would be important determinants of sensitivity. Differences in potentiation of MO toxicity were observed between mice expressing $hPON1_{Q192}$ and $hPON1_{R192}$ with pre-exposure to CPO, but not DZO or PO.

The differences in genotype-modulation of potentiation among OP compounds are consistent with the catalytic efficiencies of the $hPON1_{Q192}$ and $hPON1_{R192}$ alloforms. Thus, PON1 status can have impacts on the detoxication of chemicals that are not direct PON1 substrates. Presumably, this modulation of OP mixture toxicity by PON1 would be relevant for not only MO, but also for other compounds that are detoxified or bioactivated by CaEs, including drugs, pro-drugs, pyrethroid insecticides, and other OP compounds (Abernathy and Casida, 1973; Gaughan et al. 1980; Godin et al. 2007; Choi et al. 2004; Wheelock et al. 2005).

Of particular relevance to the toxicity of OP mixtures is a study by Lu et al. (2006) that measured pesticide metabolites in the urine of children, with the most commonly-occurring metabolites being malathion dicarboxylic acid (MDA, a metabolite of malathion), and 3,5,6-trichloro-2-pyridinol (TCPY, a metabolite of chlorpyrifos). Newborns have very low levels of PON1 (Cole et al. 2003), and would be particularly susceptible to the interactive toxicity of OP mixtures. In the case of co-exposure to malathion and chlorpyrifos, children homozygous for $hPON1_{Q192}$ represent a particularly susceptible population for both AChE inhibition by chlorpyrifos and for the interactive effects on malathion toxicity. Children of farm workers face additional risk due to multiple pathways of exposure and proximity to sources of OPs (Fenske et al. 2005).

Acknowledgments The authors thank Drs. Diana Shih, Aldons J. Lusis, and Aaron Tward for providing the $PON1^{-/-}$ mice and the $hPON1_{Q192}$ and $hPON1_{R192}$ transgenic mice used in this study. This work was supported by National Institutes of Health Grants ES09883, ES04696, ES07033, and ES09601/EPA: RD-83170901. Figures were reproduced from a previously published manuscript (Jansen et al. 2009), with permission from Elsevier Press.

References

Abernathy CO, Casida JE (1973) Pyrethroid insecticides: esterase cleavage in relation to selective toxicity. Science 179:1235–1236

Adkins S, Gan KN, Mody M, La Du BN (1993) Molecular basis for the polymorphic forms of human serum paraoxonase/arylesterase: glutamine or arginine at position 191, for the respective A or B allozymes. Am J Hum Genet 53:598–608

Aldridge WN (1954) Tricresyl phosphates and cholinesterase. Biochem J 56:185–189

Buratti FM, Testai E (2005) Malathion detoxication by human hepatic carboxylesterase and its inhibition by isomalathion and other pesticides. J Biochem Mol Toxicol 19:406–414

Casida JE, Eto M, Baron RL (1961) Biological activity of a tri-o-cresyl phosphate metabolite. Nature 191:1396–1397

Casida JE, Baron RL, Eto M, Engel JL (1963) Potentiation and neurotoxicity induced by certain organophosphates. Biochem Pharmacol 12:73–83

Chambers HW, Brown B, Chambers JE (1990) Noncatalytic detoxication of six organophosphorus compounds by rat liver homogenates. Pestic Biochem Physiol 36:308–315

Chambers JE, Ma T, Boone JS, Chambers HW (1994) Role of detoxication pathways in acute toxicity levels of phosphorothionate insecticides in the rat. Life Sci 54:1357–1364

Chen PR, Tucker WP, Dauterman WC (1969) Structure of biologically produced malathion monoacid. J Agr Food Chem 17:86–90

Choi J, Hodgson E, Rose RL (2004) Inhibition of transpermethrin hydrolysis in human liver fractions by chloropyrifos oxon and carbaryl. Drug Metabol Drug Interact 20:233–246

Cohen SD, Murphy SD (1971a) Malathion potentiation and inhibition of hydrolysis of various carboxylic esters by triorthotolyl phosphate (TOTP) in mice. Biochem Pharmacol 20:575–587

Cohen SD, Murphy SD (1971b) Carboxylesterase inhibition as an indicator of malathion potentiation in mice. J Pharmacol Exp Ther 176:733–742

Cohen SD, Callaghan JE, Murphy SD (1972) Investigation of multiple mechanisms for potentiation of malaoxon's anticholinesterase action by triorthotolyl phosphate. Proc Soc Exp Biol Med 141:906–910

Cole TB, Jampsa RL, Walter BJ, Arndt TA, Richter RJ, Shih DM, Tward A, Lusis AJ, Jack RM, Costa LG, Furlong CE (2003) Expression of human paraoxonase (PON1) during development. Pharmacogenetics 13:357–364

Cole TB, Walter BJ, Shih DM, Tward AD, Lusis AJ, Timchalk C, Richter RJ, Costa LG (2005) Toxicity of chlorpyrifos and chlorpyrifos oxon in a transgenic mouse model of the human paraoxonase (PON1) Q192R polymorphism. Pharmacogenet Genomics 15:589–598.

Cook JW, Blake JR, Williams MW(1957) Paraoxonase 1 (PON1) modulates the toxicity of mixed organophosphorus compounds. J Assess Office Agr Chem 40:664

Cook JW, Yip G (1958) Malathionase. II. Identity of a malathion metabolite. J Assess Office Agr Chem 41:407–411

Costa LG, McDonald BE, Murphy SD, Omenn GS, Richter RJ, Motulsky AG, Furlong CE (1990) Serum paraoxonase and its influence on paraoxon and chlorpyrifos-oxon toxicity in rats. Toxicol Appl Pharmacol 103:66–76

Costa LG, Li W-F, Richter RJ, Shih DM, Lusis AJ, Furlong CE (1999) The role of paraoxonase (PON1) in the detoxication of organophosphates and its human polymorphism. Chem Biol Interact 119–120:429–438

Costa LG (2006) Current issues in organophosphate toxicology. Clin Chim Acta 366:1–13

Davies H, Richter RJ, Kiefer M, Broomfield C, Sowalla J, Furlong CE (1996) The human serum paraoxonase polymorphism is reversed with diazinon, soman and sarin. Nat Genet 14: 334–336.

DuBois KP, Doull J, Deroin J, Cumming OR (1953) Studies on the toxicity and mechanism of action of some new insecticidal thionophosphates. Arch Ind Hyg Occup Med 8:350–358

DuBois KP (1958) Potentiation of toxicity of insecticidal organophosphates. Arch Industr Health 18:488–496

DuBois KP (1969) Combined effects of pesticides. Canad Med Assoc J 100:173–179

Ellman GL, Courtney KD, Andres VJ, Featherstone RM (1961) A new and rapid colorimetric determination of acetylcholinesterase activity. Biochem Pharmacol 7:88–95

Eskenazi B, Rosas LG, Marks AR, Bradman A, Harley K, Holland N, Johnson C, Fenster L, Barr DB (2008) Pesticide toxicity and the developing brain. Basic Clin Pharmacol Toxicol 102:228–236

Fenske RA, Lu C, Curl CL, Shirai JH, Kissel JC (2005) Biologic monitoring to characterize organophosphorus pesticide exposure among children and workers: an analysis of recent studies in Washington State. Environ Health Perspect 113:1651–1657

Furlong CE, Richter RJ, Seidel SL, Costa LG, Motulsky AG (1989) Spectrophotometric assays fro the enzymatic hydrolysis of the active metabolites of chlorpyrifos and parathion by plasma paraoxonase/arylesterase. Anal Biochem 180:242–247

Furlong CE, Costa LG, Hassett C, Richter RJ, Sundstrom JA, Adler DA, Disteche CM, Omiecinski CJ, Chapline C, Crabb JW (1993) Human and rabbit paraoxonases: purification, cloning, sequencing, mapping and role of polymorphism in organophosphate detoxification. Chem Biol Interact 87:35–48

Furlong CE, Holland N, Richter RJ, Bradman A, Ho A, Eskenazi B (2006) PON1 status of farmworker mothers and children as a predictor of organophosphate sensitivity. Pharmacogenet Genomics 16:183–190

Furlong CE (2007) Genetic variability in the cytochrome P450-paraoxonase 1 (PON1) pathway for detoxication of organophosphorus compounds. J Biochem Mol Toxicol 21:197–205

Furlong CE, Richter RJ, Li W-F, Brophy VH, Carlson C, Meider M, Nickerson D, Costa LG, Ranchalis J, Lusis AJ, Shih DM, Tward A, Jarvik GP (2008) The functional consequences of polymorphisms in the human PON1 gene. In: Mackness B, Mackness M, Aviram M, Paragh G (Eds). The Paraoxonases: Their Role in Disease, Development and Xenobiotic Metabolism. Dordrecht, The Netherlands: Springer, pp. 267–281

Gaughan LC, Engel JL, Casida JE (1980) Pesticide interactions: effects of organophosphorus pesticides on the metabolism, toxicity, and persistence of selected pyrethroid insecticides. Pestic Biochem Physiol 14:81–85

Godin SJ, Crow JA, Scollon EJ, Hughes MF, DeVito MJ, Ross MK (2007) Identification of rat and human cytochrome P450 isoforms and a rat serum esterase that metabolize the pyrethroid insecticides Deltamethrin and Esfenvalerate. Drug Metab Dispos 35:1664–1671

Hassett C, Richter RJ, Humbert R, Chapline C, Crabb JW, Omiecinski CJ, Furlong CE (1991) Characterization of cDNA clones encoding rabbit and human serum paraoxonase: The mature protein retains its signal sequence. Biochemistry 30:10141–10149

Hosokawa M, Endo T, Fujisawa M, Hara S, Iwata N, Sato Y, Satoh T (1995) Interindividual variation in carboxylesterase levels in human liver microsomes. Drug Metab Dispos 23:1022–1027

Huen K, Richter R, Furlong C, Eskenazi B, Holland N (2009) Validation of PON1 enzyme activity assays for longitudinal studies. Clin Chim Acta 402(1–2):67–74

Humbert R, Adler DA, Disteche CM, Hassett C, Omiecinski CJ, Furlong CE (1993) The molecular basis of the human serum paraoxonase activity polymorphism. Nat Genet 3:73–76

Jansen KL, Cole TB, Park S, Furlong CE, Costa LG (2009) Paraoxonase 1 (PON1) modulates the toxicity of mixed organophosphorus compounds. Toxicol Appl Pharmacol 236(2):142–153

Jarvik GP, Jampsa R, Richter RJ, Carlson CS, Rieder MJ, Nickerson DA, Furlong CE (2003) Novel paraoxonase (PON1) nonsense and missense mutations predicted by functional genomic assay of PON1 status. Pharmacogenetics 13:291–295

Li WF, Costa LG, Furlong CE (1993) Serum paraoxonase status: a major factor in determining resistance to organophosphates. J Toxicol Environ Health 40:337–346

Li WF, Furlong CE, Costa LG (1995) Paraoxonase protects against chlorpyrifos toxicity in mice. Toxicol Lett 76:219–226

Li WF, Costa LG, Richter RJ, Hagen T, Shih DM, Tward A, Lusis AJ, Furlong CE (2000) Catalytic efficiency determines the *in vivo* efficacy of PON1 for detoxifying organophosphates. Pharmacogenetics 10:767–799

Li B, Sedlacek M, Manoharan I, Boopathy R, Duysen EG, Masson P, Lockridge O (2005) Butyrylcholinesterase, paraoxonase, and albumin esterase, but not carboxylesterase, are present in human plasma. Biochem Pharmacol 70:1673–1684

Lu C, Toepel K, Irish R, Fenske RA, Barr DB, Bravo R (2006) Organic diets significantly lower children's dietary exposure to organophosphorus pesticides. Environ Health Perspect 114: 260–263

Main AR (1956) The role of A-esterase in the acute toxicity of paraoxon, TEPP and parathion. Can J Biochem Physiol 34:197–216

Main AR, Dauterman WC (1967) Kinetic for the inhibition of carboxylesterase by malaoxon. Can J Biochem 45:757–771

March RB, Fukuto TR, Metcalf RL, Moxon MG (1956) Fate of P32 labelled malathion in the laying hen, white mouse and American cockroach. J Econ Entomol 49:185–195

Moser VC, Casey M, Hamm A, Carter WH Jr, Simmons JE, Gennings C (2005) Neurotoxicological and statistical analyses of a mixture of five organophosphorus pesticides using a ray design. Toxicol Sci 86:101–115

Moser VC, Simmons JE, Gennings C (2006) Neurotoxicological interactions of a five-pesticide mixture in preweanling rats. Toxicol Sci 92:235–245

Munger JS, Shi GP, Mark EA, Chin DT, Gerard C, Chapman HA (1991) A serine esterase released by human alveolar macrophages is closely related to liver microsomal carboxylesterases. J Biol Chem 266:18832–18838

Murphy SD, DuBois KP (1957) Quantitative measurement of inhibition of the enzymatic detoxification of malathion by EPN (ethyl p-nitrophenyl thionobenzene phosphate). Proc Soc Exp Biol Med 96:813–818

Murphy SD, Anderson RL, DuBois KP (1959) Potentiation of toxicity of malathion by triorthotolyl phosphate. Proc Soc Exp Biol Med 100:483–487

O'Brien RD (1957) Properties and metabolism in the cockroach and mouse of malathion and malaoxon. J Econ Entomol 50:1159–1164

Pond AL, Chambers HW, Coyne CP, Chambers JE (1998) Purification of two rat hepatic proteins with A-esterase activity toward chlorpyrifos-oxon and paraoxon. J Pharmacol Exp Ther 286:1404–1411

Ramakrishna N, Ramachandran BV (1978) Malathion A and B esterases of mouse liver—III: *In vivo* effect of parathion and related PNP-containing insecticides on esterase inhibition and potentiation of malathion toxicity. Biochem Pharmacol 27:2049–2054

Richter RJ, Furlong CE (1999) Determination of paraoxonase (PON1) status requires more than genotyping. Pharmacogenetics 9:745–753

Richter RJ, Jarvik GP, Furlong CE (2008) Determination of paraoxonase 1 status without the use of toxic organophosphate substrates. Circ Cardiovasc Genet 1:147–152

Richter RJ, Jarvik GP, Furlong CE (2009) Paraoxonase 1 (PON1) status and substrate hydrolysis. Toxicol Appl Pharmacol 235(1):1–9

Satoh T, Hosokawa M (2006) Structure, function and regulation of carboxylesterases. Chem Biol Interact 162:195–211

Seume FW, O'Brien RD (1960) Potentiation of toxicity to insects and mice of phosphorothionates containing carboxyester and carboxyamide groups. Toxicol Appl Pharmacol 2: 495–503

Shih DM, Gu L, Xia YR, Navab M, Li WF, Hama S, Castellani LW, Furlong CE, Costa LG, Fogelman AM, Lusis AJ (1998) Mice lacking serum paraoxonase are susceptible to organophosphate toxicity and atherosclerosis. Nature 394:284–287

Stevens RC, Suzuki SM, Cole TB, Park SS, Richter RJ, Furlong CE (2008) Engineered recombinant human paraoxonase 1 (rHuPON1) purified from Escherichia coli protects against organophosphate poisoning. Proc Natl Acad Sci USA 105:12780–12784

Su MQ, Kinoshita FK, Frawley JP, DuBois KP (1971) Comparative inhibition of aliesterases and cholinesterase in rats fed eighteen organophosphorus insecticides. Toxicol Appl Pharmacol 20:241–249

Tang J, Cao Y, Rose RL, Brimfield AA, Dai D, Goldstein JA, Hodgson E (2001) Metabolism of chlorpyrifos by human cytochrome P450 isoforms and human, mouse, and rat liver microsomes. Drug Metab Dispos 29:1201–1204

Tang J, Chambers JE (1999) Detoxication of paraoxon by rat liver homogenate and serum carboxylesterases and A-esterases. J Biochem Mol Toxicol 13:261–268

Timchalk C, Poet TS, Hinman MN, Busby AL, Kousba AA (2005) Pharmacokinetic and pharmacodynamic interaction for a binary mixture of chlorpyrifos and diazinon in the rat. Toxicol Appl Pharmacol 205:31–42

US Environmental Protection Agency (1999) Policy on a Common Mechanism of Action: The Organophosphate Pesticides. Fed Regist 64(24):5795–5799

US Environmental Protection Agency (2002) Organophosphate pesticides: Revised OP cumulative risk assessment. www.epa.gov/pesticides/cumulative/rra-op/

US Environmental Protection Agency (2006) Organophosphorus Cumulative Risk Assessment-2006 Update. Technical Executive Summary. US EPA Office of Pesticide Programs. www.epa.gov/oppsrrd1/cumulative/2006-op/

Verschoyle RD, Reiner E, Bailey E, Aldridge WN (1982) Dimethylphosphorothioates. Reaction with malathion and effect on malathion toxicity. Arch Toxicol 49:293–301

Wheelock CE, Eder KJ, Werner I, Huang H, Jones PD, Brammell BF, Elskus AA, Hammock BD (2005) Individual variability in esterase acticity and CYP1A levels in Chinook salmon (Oncorhynchus tshawytscha) exposed to esfenvalerate and chlorpyrifos. Aquat Toxicol 74:172–192

Williams FM, Mutch EM, Nicholson E, Wynne E, Wright P, Lambert D, Rawlins MD (1989) Human liver and plasma aspirin esterase. J Pharm Pharmacol 41:407–409

Winder C, Balouet JC (2002) The toxicity of commercial jet oils. Environ Res 89:146–164

Identification and Characterization of Biomarkers of Organophosphorus Exposures in Humans

Jerry H. Kim, Richard C. Stevens, Michael J. MacCoss,
David R. Goodlett, Alex Scherl, Rebecca J. Richter,
Stephanie M. Suzuki, and Clement E. Furlong

Abstract Over 1 billion pounds of organophosphorus (OP) chemicals are manu-factured worldwide each year, including 70 million pounds of pesticides sprayed in the US. Current methods to monitor environmental and occupational exposures to OPs such as chlorpyrifos (CPS) have limitations, including low specificity and sensitivity, and short time windows for detection. Biomarkers for the OP tricresyl phosphate (TCP), which can contaminate bleed air from jet engines and cause an occupational exposure of commercial airline pilots, crewmembers and passengers, have not been identified.

The aim of our work has been to identify, purify, and characterize new biomark-ers of OP exposure. Butyrylcholinesterase (BChE) inhibition has been a stan-dard for monitoring OP exposure. By identifying and characterizing molecular biomarkers with longer half-lives, we should be able to clinically detect TCP and OP insecticide exposure after longer durations of time than are currently possible.

Acylpeptide hydrolase (APH) is a red blood cell (RBC) cytosolic serine proteinase that removes *N*-acetylated amino acids from peptides and cleaves oxidized proteins. Due to its properties, it is an excellent candidate for a biomarker of exposure. We have been able to purify APH and detect inhibition by both CPS and metabolites of TCP. The 120-day lifetime of the RBC offers a much longer window for detecting exposure. The OP-modified serine conjugate in the active site tryptic peptide has been characterized by mass spectrometry.

This research uses functional proteomics and enzyme activities to identify and characterize useful biomarkers of neurotoxic environmental and occupational OP exposures.

Keywords Butyrylcholinesterase · Acylpeptide hydrolase · Biomarkers of OP exposure · Mass spectrometry · Affinity purification · Immunomagnetic beads

C.E. Furlong (✉)
Department of Medicine and Genome Sciences, University of Washington, Seattle, WA, USA
e-mail: clem@u.washington.edu

S.T. Reddy (ed.), *Paraoxonases in Inflammation, Infection, and Toxicology*, Advances in Experimental Medicine and Biology 660, DOI 10.1007/978-1-60761-350-3_7,
© Humana Press, a part of Springer Science+Business Media, LLC 2010

1 Introduction

Biomarkers are used in many aspects of health surveillance to identify disease presence, track disease progression, monitor drug delivery or metabolism, or monitor chemical exposure. There is a recognized need to expand and improve biomarker identification and quantification. Exposures to the xenobiotic organophosphates (OPs) range from low-level, chronic exposure during pesticide application (e.g., on farms, in residences, or in the workplace) to high-dose, acute exposures including release of nerve agents or toxic industrial OPs. OPs can have both rapid and chronic toxicity, due to their action on specific esterases and lipases, most notably acetylcholinesterase (AChE) and neuropathic target esterase (NTE). In each of the above examples, a rapid and accurate assessment of the OP to which the person was exposed, degree of exposure, and time period of exposure will help direct therapy to the victim and assess the level of threat to others. Traditionally, metabolites for specific OPs can be measured in the urine after an exposure, such as 3,5,6-trichloro-2-pyridinol for chlorpyrifos (CPS). However, there are several problems with this method. Usually, the metabolite is excreted for only a short period of several days after a significant exposure. In addition, due to the widespread use of CPS and environmental persistence of its breakdown products, the metabolites appear at a high background level in the general population (Hill et al. 1995). Their appearance does not indicate whether the individual was exposed to the harmful parent compound, or the harmless breakdown product.

Human plasma can serve as an ideal source for biomarkers of exposure, due to the ease of sample collection and the wide range of proteins held within the plasma compartment (Anderson and Anderson, 1977). In terms of monitoring OP exposure, the esterases present in plasma are ideal targets for assessing their inhibition and modification. Initial attempts at biomarker discovery focused on the major plasma esterases, including AChE, butyrylcholinesterase (BChE), and albumin (Black et al. 1999; Peeples et al. 2005). AChE is present in only trace amounts in plasma (Li et al. 2005), but also exists as membrane-bound protein on red blood cells (RBCs), where its function is unknown. In humans, measurement of RBC AChE inhibition has been one standard method of detecting OP exposure (Holmstedt, 1959), with the caveat that inhibition of AChE on the RBCs usually overestimates inhibition of AChE in the central nervous system (CNS) depending on the pharmacokinetics of the agent and passage of time since exposure. In addition, there are other difficulties with this method. The first difficulty is due to inter-individual variability. While the intra-individual coefficient of variation (CV) is about 10%, inter-individual CV is between 10 and 40% (Lotti, 1995). Measuring the individual's pre-exposure activity levels would improve precision, but this is only practical in certain situations, such as monitoring agricultural worker exposure during the growing season. Pre-exposure activity levels are rarely available for cases of non-agricultural exposure. Second, measuring AChE activity levels from RBCs will not identify the specific inhibitory OP agent.

Many animals, including pig and rodent species, have abundant carboxylesterase (CE) in plasma. In humans, CE is not free in plasma, but can be purified from a membrane-bound form on monocytes (Saboori and Newcombe, 1990). While perhaps not an ideal biomarker, CE can be inhibited by an array of OPs, including the active metabolites of CPS and tricresyl phosphate (TCP), two OPs we have been interested in.

2 Modification of Blood and Plasma Biomarkers

Owing to its freely soluble presence in plasma, BChE has been a popular target for biomarker research. Newer methods have been developed that can identify and quantify OP exposure based on BChE modification. Polhuijs and colleagues developed a novel method involving fluoride ions. When incubated with 2 M potassium fluoride at pH 4, inhibited BChE released the inhibiting OP and yielded free enzyme plus the phosphofluoridate form of the OP (Polhuijs et al. 1997). In this case, sarin was regenerated and was analyzed using gas chromatography/mass spectrometry (GC/MS). A further advance was made analyzing peptic digests of uninhibited and inhibited BChE with electrospray liquid chromatography/tandem mass spectrometry (LC/MS/MS) (Fidder et al. 2002). Using this method, they could detect methylphosphonic acid residues adducted to the active site serine, at position 192. They repeated the experiments with other OPs, including more commonly-used pesticides. In addition to this work on BChE, highly sensitive and specific MS peptide capture methods have been developed using magnetic beads (Whiteaker et al. 2007). Their protocol involved digesting the biomarker of interest with trypsin, incubating the target peptides with beads coupled to anti-peptide antibody, then eluting and analyzing the captured peptides with LC/MS/MS. The sample could be spiked with a known amount of a stable isotope of biomarker peptide or protein (utilizing ^{13}C and ^{15}N), which provided an extra peak of known magnitude next to the desired sample and appeared as a doublet. This rapid and precise enrichment can enable detection down to ng/mL concentrations.

OP modification of BChE coupled with MS detection was a major advance in diagnosing OP exposure, especially in nerve agent attacks. However, BChE has somewhat limited utility because of its 11-day half-life in plasma. Another serine esterase, acylpeptide hydrolase (APH), was proposed in 2000 as both a diagnostic and therapeutic target for OPs (Richards et al. 2000). APH was first isolated in the liver (Tsunasawa et al. 1975; Gade and Brown, 1978), then in the circulating RBC cytosol and the brain (Yamin et al. 2007). In the RBC cytosol, APH removes N-acetyl amino acids from the ends of peptides (Fujino et al. 2000). APH serves a critical function in protecting the cytosol from denatured and oxidized proteins (Shimizu et al. 2004), and loss of this function by inhibition is lethal (Yamaguchi et al. 1999). Due to its presence in the RBC, which has a lifespan of 120 days and no protein synthetic capability, OP-modified APH should be measurable for several weeks, depending on the level of exposure. Casida and colleagues characterized the inhibition profile of APH using a large array of OPs (Quistad et al. 2005). They administered sufficient di-isopropyl fluorophosphate intraperitoneally into mice to inhibit RBC APH by 100% and plasma BChE by 80% at 4 hours post-injection, and brain AChE by 40% at 8 hours post-injection. While the BChE activity returned to baseline by day 4 post-injection, only 20% of the APH activity returned at this time, with no further acitivty measurements obtained. These data suggest that APH inhibition and modification should be detectable significantly longer post-exposure than OP-modified plasma BChE.

Many other secondary targets of OPs have been identified and could also serve as biomarker candidates (Casida and Quistad, 2004). One biomarker property that we have sought is detectable inhibition by TCP metabolites. TCP is of important historical interest, since it caused paralysis afflicting tens of thousands of Americans during Prohibition (Parascandola, 1995). Many over-the-counter medicinals were tinctures, or alcoholic extracts of certain plant parts, such as leaves and roots. Unfortunately, many adulterants were added to either dilute the preparation or improve the taste. One popularly abused tincture was Jamaica ginger, or jake, which had a 70% alcohol content. TCP was used as an adulterant in the manufacture of a popular brand, since it was cheap, soluble in alcohol, and improved the strong taste of ginger.

In the early 1930s, the National Institutes of Health, then a division of the Public Health Service, identified the ortho isomer of TCP, tri-orthocresyl phosphate (TOCP), as the cause of the neural toxicity, including wrist drop and foot drop (Parascandola, 1995). The latter inspired many songs and stories of the era involving jake leg or jake walk blues. In 1954, Aldridge recognized that TOCP itself was a poor inhibitor of cholinesterases, but became a much more potent inhibitor after incubation with rat liver (Aldridge, 1954). Later, Casida and colleagues determined that the active metabolite responsible for the toxicity of TOCP was saligenin cyclic-*o*-tolyl phosphate (Casida et al. 1961). Phenyl saligenin phosphate (PSP) is an analogue of this metabolite, which we use in its place since it is easier to obtain. TCP is currently of interest because of its continued use as a plasticizer and an additive in jet-engine lubricants and hydraulic fluids. In the US, over 20 million pounds of TCP is used annually, and it is a concern due to continued occupational exposure during its manufacture and use, despite its known toxicity (Winder and Balouet, 2002). Hundreds of crewmembers have reported exposure to cabin fumes, resulting in memory loss and cognitive dysfunction of sufficient severity to result in lost time at work and even permanent removal from the workforce. To date, there have been no published biomarkers that enable detection beyond several days after exposure to TCP or TOCP. One focus of our research is to identify biomarkers that can identify common pesticide exposure as well as TCP exposure.

3 Identification and Characterization of Biomarkers of OP Exposure from Human Blood

To test the feasibility of identifying peptides modified by triaryl phosphates, we inhibited pig carboxylesterase (PCE) using a mixture of TCP isomers that inhibit PCE without bioactivation. After incubating with TCP, modified (inhibited) PCE was digested with trypsin and the peptides were analyzed by tandem mass spectrometry on a linear ion trap, with multidimensional protein identification technology (MudPIT). The tryptic digests were loaded onto a nanoflow chromatography column and then eluted using a reverse-phase gradient. The effluent was electrosprayed into the mass spectrometer and the eluting peptides were detected and fragmented automatically by data-dependent acquisition. The resulting MS/MS spectra provided evidence that identified modified aged and un-aged peptide residues of PCE fragments. The aged peptide modifications resulted from the loss of one cresyl group thus generating a negative charge, a necessity for the neurotoxicity that is associated with TCP. This was identified as a 170-Da shift of the peptide containing the active site serine residue, 222 (Fig. 1) (Furlong et al. 2005).

We next sought to apply this approach to human plasma BChE. Our first aim was to generate sufficient human plasma BChE for the required experiments. BChE has been isolated from human plasma by a variety of methods (Lockridge and La Du, 1978; Ralston et al. 1983; Grunwald et al. 1997; Mehrani, 2004; Lockridge et al. 2005). Most recent methods have used the procainamide affinity column chromatographic technique first described by Lockridge and La Du in 1978. We attempted to simplify the methods using more available anion exchange and hydrophobic interaction chromatography (HIC). In addition, since our laboratory routinely processes quantities of plasma for purification of paraoxonase 1 (PON1), we wanted to design a protocol for purifying both proteins from the same starting material. De-identified plasma was treated with Cibacron Blue Agarose to separate the HDL-associated proteins from BChE. The eluted BChE fractions were pooled and further resolved with ion exchange chromatography. Consecutive octyl-HIC columns were used for further purification before a final purification step with Superdex gel filtration column. Figure 2 shows an SDS-PAGE analysis of the purification steps.

The purified BChE was used to raise antibodies for rapid processing of subject samples. These rapid protocols will utilize immunomagnetic bead (IMB) separation procedures. Further rapid purification protocols will make use of IMB procedures. In preliminary studies, we linked tosyl-activated magnetic beads (Dynal Dynabeads, Invitrogen) with commercial human anti-BChE antibody. After separating the plasma fraction from a human whole blood sample, we added several OPs to inhibit BChE activity, then incubated both inhibited and uninhibited plasma samples with magnetic beads. Figure 3a shows a 93% depletion of BChE activity following incubation of plasma with IMB. Figure 3b shows BChE inhibition by two different OPs, PSP and chlorpyrifos oxon (CPO). After BChE was bound to the beads, the beads were magnetically separated and eluted in formic acid. The BChE appears as the dark band in the silver-stained gel (Fig. 3c).

The next biomarker candidate that we purified and characterized was the RBC acylpeptide hydrolase (APH). Vose et al. (2007) characterized an RBC lysophosphatidylcholine hydrolase as a promising candidate for identifying OP-delayed neurotoxicants. We designed a protocol that would purify a soluble lipase from RBCs. Using a combination of hydrophobic interaction chromatography (HIC) and anion exchange chromatography (DEAE), we purified an RBC esterase to near homogeneity (20,000 fold) (Fig. 4a). Figure 4b shows an SDS-PAGE silver-stain

```
Ser63      AVFLGVPFAKPPLGS*LRFAP
           FLGVPFAKPPLGS*L

Ser222     GGDPGSVTIFGES*AGGESVSVL
                  TIFGES*AGGESV
                  TIFGES*AGGESVS
                   IFGES*AGGESVSV
                   IFGES*AGGESVSVL

Ser368     IVGINKQEFGWLLPTM@GFPLS*EG
                QEFGWLLPTM@GFPLS*E

Ser384     KS*YPIANIPEELTPVATDKY

Ser473     SSDKKPKTVIGDFGDEIFS*VFGFPLLKGDA
                   GDFGDEIFS*VFGFPLLKGDAPEEEV
                   GDFGDEIFS*VFGFPLLKGDAPEEEVS
                   GDFGDEIFS*VFGFPLLKGDAPEEEVSLS
                    DFGDEIFS*VFGFPLLKGDAPEEEV
                    DFGDEIFS*VFGFPLLKGDAPEEEVS
                    DFGDEIFS*VFGFPLLKGDAPEEEVSLS
                        FS*VFGFPLLKGDAPEEEVSLSK
```

Fig. 1 Modified porcine carboxylesterase (PCE) peptides. Aged peptides identified with a mass shift of 170 Da on serine are indicated by S*. Ser222 is the catalytic serine, with Ser368, Ser384, and Ser473 located nearby. (Adapted with permission from Fig. 2 in Furlong et al. 2005)

Fig. 2 SDS-PAGE analysis of the purification steps of butyrylcholinesterase (BChE). (**a**) Original whole plasma, (**b**) Flow-through after Cibacron Blue, (**c**) Molecular weight markers (listed in kDa), (**d**) Active BChE fractions after two octyl-HIC columns, (**e**) Active BChE fractions after Superdex 200, (**f**) Our internal BChE standard. *Arrow* indicates BChE mobility

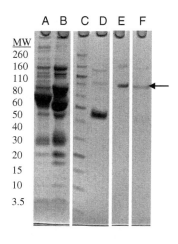

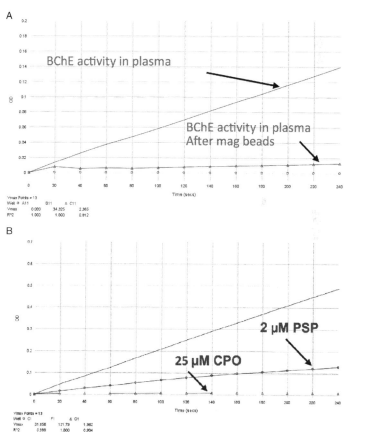

Fig. 3 (**a**) Removal of butyrylcholinesterase (BChE) after incubation of human plasma with magnetic beads bound to anti-BChE antibody. (**b**) BChE inhibition after incubating human plasma with 2 μM phenyl saligenin phosphate (PSP) and 25 μM chlorpyrifos oxon (CPO). (**c**) Purified BChE (*arrow*), after magnetic bead capture, elution, and analysis by silver staining of SDS-PAGE

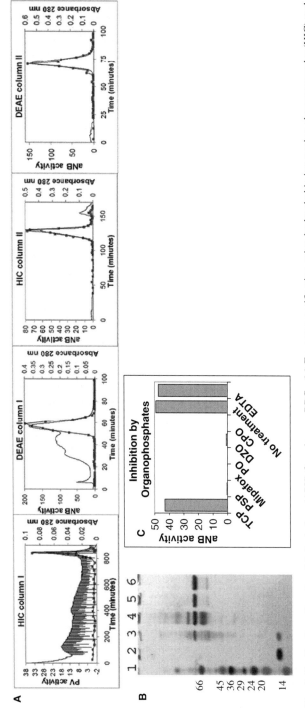

Fig. 4 Purification of acylpeptide hydrolase (APH) from human RBCs. (**a**) Four-step purification using hydrophobic interaction chromatography (HIC) and anion exchange chromatography (DEAE) resulting in >95% purity. (**b**) Silverstained SDS-PAGE analysis of purification steps; lane 1, molecular weight markers; lane 2, RBC extract; lane 3, HIC column I; lane 4, DEAE column I; lane 5, HIC column II; and lane 6, DEAE column II. (**c**) Inhibition by 25–40 μM OP compounds

Fig. 5 The sequence of the acylpeptide hydrolase (APH) active site peptide with the active site serine (587) after tryptic digestion and modification with phenyl saligenin phosphate (PSP)

156 Da

|

GGS*HGGFISCHLIGQYPETY

analysis of each step of the purification procedure. Purification of this esterase has been carried out with samples as small as one millilitre of RBC lysate. The purified esterase was further gel-purified, excised, and digested with trypsin. Peptides were separated and characterized by reverse phase LC/MS/MS. The esterase was identified as acetyl peptide hydrolase (APH) by mass spectrometry. Quistad et al. (2005) had previously proposed APH as a sensitive biomarker for exposure to specific OP compounds. To determine the inhibitor specificity of the purified APH, we incubated it with a variety of OPs, including TCP and PSP. The purified APH was inhibited by CPO, diazoxon (DZO), paraoxon (PO), PSP, and the classical NTE inhibitor, mipafox, but not TCP (Fig. 4c). However, since APH was inhibited by PSP, an analogue of the active TOCP metabolite generated in vivo by hepatic transformation, we suspect that a TCP exposure containing any of the ortho analogs (mono-, di- or tri-orthocresyl phosphate) will result in APH inhibition in actual cases of exposure. The identity of the aged OP on the active site serine of APH was verified by mass spectrometry (Fig. 5).

4 Conclusions

Expanding the identification and characterization of biomarkers beyond BChE is necessary for detecting and treating poisonous OP exposure. The approaches described here have included the standard biomarker BChE, as well as several others such as APH. The approach involves the development of rapid protocols for extraction of the target biomarker protein from a sample, digesting with the appropriate enzyme and identifying the OP-modified peptide by mass spectrometry. Additional directions that have been underway include expressing recombinant, active biomarker proteins in an *E. coli* system to provide heavy isotope-labeled standards to use in quantifying the degree of modification. We feel these methods are optimal for filling a void of diagnosing and treating long-term exposures to several ubiquitous OPs.

Acknowledgments This work was supported by NIH Grants R01ES09883, P42ES04696, and NIEHS P30ES07033 (UW CEEH), as well as funding from Pilot Unions, Flight Attendant Unions, the Royal Australian Air Force, the Norwegian Union of Energy Workers (SAFE) and NYCO S.A. We thank Dr. Marian Ehrich for the kind gift of PSP, and Dr. Oksana Lockridge for BChE and anti-BChE antibodies.

References

Aldridge WN (1954) Tricresyl phosphates and cholinesterase. Biochem J 56(2):185–189

Anderson L, Anderson NG (1977) High resolution two-dimensional electrophoresis of human plasma proteins. Proc Natl Acad Sci USA 74(12):5421–5425

Black RM, Harrison JM, Read RW (1999) The interaction of sarin and soman with plasma proteins: the identification of a novel phosphonylation site. Arch Toxicol 73(2):123–126

Casida JE, Eto M, Baron RL (1961) Biological activity of a tri-o-cresyl phosphate metabolite. Nature 191:1396–1397

Casida JE, Quistad GB (2004) Organophosphate toxicology: safety aspects of nonacetyl-cholinesterase secondary targets. Chem Res Toxicol 17(8):983–998

Fidder A, Hulst AG, Noort D, de Ruiter R, van der Schans MJ, Benschop HP, Lanenberg JP (2002) Retrospective detection of exposure to organophosphorus anti-cholinesterases: mass spectrometric analysis of phosphylated human butyrylcholinesterase. Chem Res Toxicol 15(4):582–590

Fujino T, Watanabe K, Beppu M, Kikugawa K, Yasuda H (2000) Identification of oxidized protein hydrolase of human erythrocytes as acylpeptide hydrolase. Biochim Biophys Acta 1478(1):102–112

Furlong CE, Cole TB, MacCoss M, Richter R, Costa LG (2005) Biomarkers for Exposure and of Sensitivity to Organophosphorus [OP] Compounds. Proceedings of the BALPA Air Safety and Cabin Air Quality International Aero Industry Conference. April 20-21. Imperial College, London.

Gade W, Brown JL (1978) Purification and partial characterization of alpha-N-acylpeptide hydrolase from bovine liver. J Biol Chem 253(14):5012–5018

Grunwald J, Marcus D, Papier Y, Raveh L, Pittel Z, Asani Y (1997) Large-scale purification and long-term stability of human butyrylcholinesterase: a potential bioscavenger drug. J Biochem Biophys Methods 34(2):123–135

Hill RH, Head SL, Baker S, Gregg M, Shealy DB, Bailey SL, Williams CC, Sampson EJ, Needham LL (1995) Pesticide residues in urine of adults living in the United States: reference range concentrations. Environ Res 71(2):99–108

Holmstedt B (1959) Pharmacology of organophosphorus cholinesterase inhibitors. Pharmacol Rev 11:567–688

Li B, Sedlacek M, Manoharan I, Boopathy R, Duysen EG, Masson P, Lockridge O (2005) Butyrylcholinesterase, paraoxonase, and albumin esterase, but not carboxylesterase, are present in human plasma. Biochem Pharmacol 70(11):1673–1684

Lockridge O, La Du B (1978) Comparison of atypical and usual human serum cholinesterase. Purification, number of active sites, substrate affinity, and turnover number. J Biol Chem 253(2):361–366

Lockridge O, Schopfer LM, Winger G, Woods JH (2005) Large scale purification of butyryl-cholinesterase from human plasma suitable for injection into monkeys; a potential new therapeutic for protection against cocaine and nerve agent toxicity. J Med Chem Biol Radiol Def 3:5095

Lotti M (1995) Cholinesterase inhibition: complexities in interpretation. Clin Chem 41:1814–1818

Mehrani H (2004) Simplified procedures for purification and stabilization of human plasma butyrylcholinesterase. Process Biochem 39(7):877–882

Parascandola J (1995) The Public Health Service and Jamaica ginger paralysis in the 1930s. Public Health Reports 110(3):361–363

Peeples ES, Schopfer LM, Duysen EG, Spaulding R, Voelker T, Thompson CM, Lockridge O (2005) Albumin, a new biomarker of organophosphorus toxicant exposure, identified by mass spectrometry. Toxicol Sci 83(2):303–312

Polhuijs M, Langenberg JP, Benschop HP (1997) New method for retrospective detection of exposure to organophosphorus anticholinesterases: application to alleged sarin victims of Japanese terrorists. Toxicol Appl Pharmacol 146(1):156–161

Quistad GB, Klintenberg R, Casida JE (2005) Blood acylpeptide hydrolase activity is a sensitive marker for exposure to some organophosphate toxicants. Toxicol Sci 86(2):291–299

Ralston JS, Main AR, Kilpathrick BF, Chasson AL (1983) Use of procainamide gels in the purification of human and horse serum cholinesterases. Biochem J 211(1):243–250

Richards PG, Johnson MK, Ray DE (2000) Identification of acylpeptide hydrolase as a sensitive site for reaction with organophosphorus compounds and a potential target for cognitive enhancing drugs. Mol Pharmacol 58(3):577–583

Saboori AM, Newcombe DS (1990) Human monocyte carboxylesterase. Purification and kinetics. J Biol Chem 265(32):19792–19799

Shimizu K, Kiuchi Y, Ando K, Hayakawa M, Kikugawa K (2004) Coordination of oxidized protein hydrolase and the proteasome in the clearance of cytotoxic denatured proteins. Biochem Biophys Res Commun 324(1):140–146

Tsunasawa S, Narita K, Ogata K (1975) Purification and properties of acylamino acid-releasing enzyme from rat liver. J Biochem 77(1):89–102

Vose SC, Holland NT, Eskenazi B, Casida JE (2007) Lysophosphatidylcholine hydrolases of human erythrocytes, lymphocytes, and brain: sensitive targets of conserved specificity for organophosphorus delayed neurotoxicants. Toxicol Appl Pharmacol 224(1):98–104

Whiteaker JR, Zhao L, Zhang HY, Feng LC, Piening BD, Anderson L, Paulovich AG (2007) Antibody-based enrichment of peptides on magnetic beads for mass-spectrometry-based quantification of serum biomarkers. Anal Biochem 362(1):44–54

Winder C, Balouet JC (2002) The toxicity of commercial jet oils. Environ Res 89(2):146–164

Yamaguchi M, Kambayashi D, Toda J, Sano T, Toyoshima S, Hojo H (1999) Acetylleucine chloromethyl ketone, an inhibitor of acylpeptide hydrolase, induces apoptosis of U937 cells. Biochem Biophys Res Commun 263(1):139–142

Yamin R, Bagchi S, Hildebrant R, Scaloni A, Widom RL, Abraham CR (2007) Acyl peptide hydrolase, a serine proteinase isolated from conditioned medium of neuroblastoma cells, degrades the amyloid-beta peptide. J Neurochem 100(2):458–467

Temporal and Tissue-Specific Patterns of *Pon3* Expression in Mouse: In situ Hybridization Analysis

Diana M. Shih, Yu-Rong Xia, Janet M. Yu and Aldons J. Lusis

Abstract *PON3* is a member of the paraoxonase gene family that includes *PON1* and *PON2*. For example, *PON3* and *PON1* share approximately 60% identity at the amino acid level. Recent studies have demonstrated that PON3 is present in human and rabbit HDL but not in mouse HDL. Mouse PON3 appears to be cell-associated and is expressed in a wide range of tissues such as liver, adipose, macrophage, and the artery wall. In vitro studies have shown that PON3 can prevent LDL oxidation and destroy bacterial quorum-sensing molecules. Previous studies also showed that human *PON3* transgenic mice were protected from obesity and atherosclerosis in both the C57BL/6J wild-type and LDLR knockout genetic background. Administration of adenovirus expressing the human *PON3* gene into apoE –/– mice also decreased atherosclerotic lesion formation. In order to further understand the functions of PON3 in physiology and disease, we performed in situ hybridization analysis to examine *Pon3* gene expression patterns in newborn and adult mice, in various tissues, including atherosclerotic lesions of apoE –/– mice. Our results show relatively high levels of *Pon3* mRNA labeling in the adrenal gland, submaxillary gland, lung, liver, adipose, pancreas, large intestine, and other tissues of newborn mice. In the adult mouse, *Pon3* mRNA levels were much lower in the corresponding tissues as mentioned above for the newborn mouse. Sections of the aortic root from the hearts of both wild-type and apoE –/– mice displayed moderate levels of *Pon3* mRNA labeling. *Pon3* mRNA was also detected in the atherosclerotic lesion areas at the aortic root of apoE –/– hearts. Our data revealed that mouse *Pon3* is expressed in a wide range of tissues, and that its expression is temporally controlled.

Keywords PON3 · In situ hybridization · Developmental regulation · Mouse

D.M. Shih (✉)
Division of Cardiology, David Geffen School of Medicine at UCLA, Los Angeles, CA, 90095-1679, USA
e-mail: dshih@mednet.ucla.edu

S.T. Reddy (ed.), *Paraoxonases in Inflammation, Infection, and Toxicology*, Advances in Experimental Medicine and Biology 660, DOI 10.1007/978-1-60761-350-3_8,
© Humana Press, a part of Springer Science+Business Media, LLC 2010

1 Introduction

The paraoxonase gene family contains 3 members, *PON1*, *PON2*, and *PON3*, located as a cluster on mouse chromosome 6 and human chromosome 7. The three human PON proteins share abouty 60% identity in amino acid sequence. Human PON1 is expressed primarily in the liver and is found associated with HDL particles in the blood (Blatter et al. 1993; Hassett et al. 1991). Human PON3 is expressed primarily in the liver, with lower expression levels seen in other tissues (Reddy et al. 2001; Shamir et al. 2005). Human PON2, on the other hand, is ubiquitously expressed and is found in a variety of tissues (Ng et al. 2001). In addition, whereas human PON1 and PON3 associate with HDL in the circulation, PON2 protein is not associated with HDL or LDL, but appears to remain intracellular, associated with membrane fractions of the cell (Ng et al. 2001). Our recent studies showed that mouse PON3, unlike human PON3, is not detectable in circulation or HDL (Ng et al. 2007; Shih et al. 2007), suggesting mouse PON3 is a cell-associated protein like PON2. A recent study also suggests that human PON1, PON2, and PON3 are localized to the endoplasmic reticulum (Rothem et al. 2007). We and others have shown that all three PON proteins exhibit lactonase activities with many common substrates (Draganov et al. 2005; Ozer et al. 2005; Yang et al. 2005). For example, all three PONs very efficiently metabolize 5-hydroxy-eicosatetraenoic acid 1,5-lactone and 4-hydroxy-docosahexaenoic acid, which are products of both enzymatic and nonenzymatic oxidation of arachidonic acid and docosahexaenoic acid, respectively, and may represent the endogenous substrates of PONs. All PONs, especially PON3, also have been shown to hydrolyze estrogen esters (Teiber et al. 2007), likely endogenous substrates for PONs. Furthermore, human and mouse PONs have been shown by us and other groups to hydrolyze and thereby inactivate bacterial quorum-sensing molecules, N-acyl-homoserine lactones, such as N-(3-oxododecanoyl)-L-homoserine lactone (3OC12-HSL) (Draganov et al. 2005; Ozer et al. 2005; Yang et al. 2005). These molecules are extracellular quorum-sensing signals secreted by Gram-negative bacteria such as the pathogenic bacteria, *Pseudomonas aeruginosa*, to regulate biofilm formation and secretion of virulence factors (Parsek and Greenberg, 2000). Our in vitro studies using PON1 knockout (KO) mouse serum (Ozer et al. 2005) and airway epithelial cells isolated from the PON2-deficient mice also demonstrated that both PON1 and PON2 are important for degradation of 3OC12-HSL (Stoltz et al. 2007). Therefore, growing evidence suggests possible roles of PONs in innate immunity against bacterial infection.

Both rabbit and human PON3 have been shown to associate with HDL in the circulation (Draganov et al. 2000; Reddy et al. 2001). Rabbit PON3, when purified from serum, inhibits copper-induced LDL oxidation in vitro (Draganov et al. 2000). Our studies have shown that LDL incubated with stably transfected cells overexpressing human PON3 has significantly less lipid hydroxides, and is less capable of inducing monocyte chemotactic activity (Reddy et al. 2001). Our studies of transgenic mice overexpressing human PON3 showed that elevated PON3 levels protect against atherosclerosis and obesity (Shih et al. 2007). Administration of adenovirus expressing the human *PON3* gene into apoE –/– mice also decreased atherosclerotic lesion formation (Ng et al. 2007). Therefore, PON3 appears to protect against atherosclerosis and metabolic disorders such as obesity. In order to further understand the functions of PON3 in physiology and disease, we performed in situ hybridization analysis to examine *Pon3* mRNA expression patterns in newborn and adult mice. Our results revealed that mouse *Pon3* is expressed in a wide range of tissues, and that *Pon3* expression is temporally controlled with higher expression levels detected in newborn mice as compared to the adult mice.

2 Methods

2.1 Tissue Fixation, Embedding, and Pretreatment

Whole body sections of C57BL/6 newborn and adult mice were used. The animals were sacrificed in a CO_2 chamber. Hearts of male apoE –/– mice that were 6 months old and maintained on a chow diet were also collected. These hearts were embedded in OCT medium (OCT Compound, 4583 Tissue-Tek). Tissues were frozen and cut into 10-micron sections, mounted on gelatin-coated slides and stored at –80°C. Before in situ hybridization (ISH), they were fixed in 4% formaldehyde (freshly made from paraformaldehyde; Sigma Aldrich P6148) in phosphate buffered saline (PBS), treated with triethanolamine/acetic anhydride, washed and dehydrated with a series of ethanol. Before proceeding to the ISH with *Pon3* probes, all tissues were validated with riboprobes to LDL receptor mRNA (data not shown).

2.2 cRNA Probe Preparation

A 631 bp mouse *Pon3* cDNA fragment cloned into the pBluescript II KS plasmid was used for generation of anti-sense and sense cRNA transcripts. The cRNA transcripts were synthesized in vitro according to manufacturer's conditions (Ambion) and labeled with ^{35}S-UTP (> 1000 Ci/mmol; Amersham).

2.3 Hybridization and Washing Procedures

Sections were hybridized overnight at 55°C in 50% deionized formamide, 0.3 M NaCl, 20 mM Tris-HCl pH 7.4, 5 mM EDTA, 10 mM NaH_2PO_4, 10% dextran sulphate, 1 × Denhardt's, 50 mg/ml total yeast RNA, and 50–80,000 cpm/ml ^{35}S-labeled cRNA probe. The tissue was subjected to stringent washing at 65°C in 50% formamide, 2 × SSC, 10 mM DTT and washed in PBS before treatment with 20 mg/ml RNAse A at 37°C for 30 min. Following washes in 2 × SSC and 0.1 × SSC for 10 min at 37°C, the slides were dehydrated, exposed to Kodak BioMaxMR

X-ray film, then dipped in Kodak NTB nuclear track emulsion and exposed in light-tight boxes with desiccant at 4°C.

2.4 Imaging

Photographic development was carried out in Kodak D-19. Slides were counterstained lightly with cresyl violet and analyzed using both brightfield and darkfield optics. Sense (control) cRNA probes (identical to the mRNAs) always gave background levels of hybridization signal.

3 Results

3.1 Pon3 mRNA Expression in Newborn Mice

Pon3 mRNA distribution in wild-type newborn (P1) mouse whole body sections was examined. X-ray film autoradiography provided evidence for the presence of relatively high Pon3 mRNA concentrations in the adipose tissue, submaxillary gland, liver, lung, pancreas, large intestine, adrenal cortex (Fig. 1), and stomach (Fig. 2). Figure 2 shows higher magnification images of liver and the adjacent stomach. High levels of Pon3 mRNA were detected in the hepatocytes of liver and in the glandular epithelium of stomach (Fig. 2). Higher magnification images following emulsion autoradiography showed high levels of Pon3 mRNA in the adipose tissue (Fig. 3), whereas skeletal muscle and bone remain unlabeled. Emulsion autoradiography detected high levels of Pon3 mRNA in the bronchiole epithelium of lung (Fig. 4).

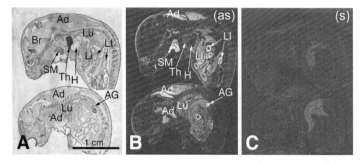

Fig. 1 Pon3 mRNA distribution in wild-type newborn (P1) mouse whole body sections. (**a**) Anatomical view of whole body sections following staining with cresyl violet seen under brightfield illumination. (**b**) X-ray film autoradiography detection of Pon3 mRNA seen as bright labeling. (**c**) Control hybridization in an adjacent section comparable to (**b**). Abbreviations: Ad – adipose tissue; AG – adrenal gland; Br – brain; H – heart; Li – liver; LI – large intestine; Lu – lung; SM – submaxillary gland; Th – thymus; (as) – antisense; (s) – sense

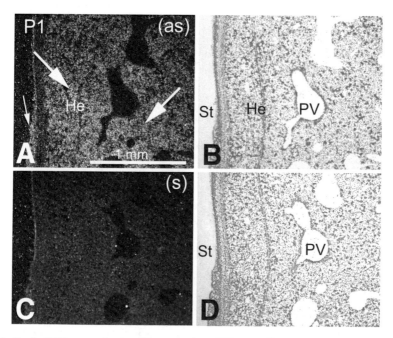

Fig. 2 *Pon3* mRNA expression in wild-type newborn (P1) mouse liver. (**a**) Emulsion autoradiography detection of *Pon3* mRNA in hepatic tissue (*arrows*) and glandular epithelium (*small arrow*) of stomach seen as bright labeling under darkfield illumination. (**b**) Cresyl violet staining of the same section shown in (**a**) seen under brightfield illumination. (**c**) Control hybridization in an adjacent section comparable to (**a**). (**d**) Cresyl violet staining of the same section shown in (**c**) seen under brightfield illumination. Abbreviations: He – hepatocyte; PV – portal vein, branch; St – stomach; (as) – antisense; (s) – sense

High levels of *Pon3* mRNA were also detected in the acinar cells of pancreas, but not in the pancreatic islet (Figs. 5 and 10a).

3.2 Pon3 *mRNA Expression in Adult Mice*

Highly sensitive X-ray film autoradiography provided evidence of *Pon3* expression in the adrenal cortex, salivary glands, adipose tissue, lung, liver, and skin of adult mouse (Fig. 6). Emulsion autoradiography confirmed the X-ray film findings. As detailed in Figs. 7, 8, 9, and 10, the levels of *Pon3* expression in adult adipose tissue, liver, lung, and pancreas appeared to be much lower when compared to their newborn mouse counterparts. Similar to the newborn (Fig. 7a), in adult mouse *Pon3* mRNA was detected in the adipose tissue but not in the adjacent skeletal muscle (Fig. 7c). As seen in the newborn pancreas (Fig. 10a), *Pon3* mRNA is only detected in the acinar cells of adult pancreas, but not in the pancreatic islets (Fig. 10c).

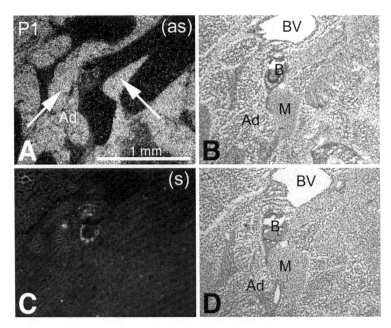

Fig. 3 *Pon3* mRNA expression in wild-type newborn (P1) mouse adipose tissue. (**a**) Emulsion autoradiography detection of *Pon3* mRNA seen as bright labeling under darkfield illumination. Labeling is seen in adipose tissue (*arrows*), whereas skeletal muscle and bone remain unlabeled. (**b**) Cresyl violet staining of the same section shown in (**a**) seen under brightfield illumination. (**c**) Control hybridization in an adjacent section comparable to (**a**). (**d**) Cresyl violet staining of the same section shown in (**c**) seen under brightfield illumination. Abbreviations: Ad – adipose tissue; B – bone; BV – blood vessel; M – skeletal muscle; (as) – antisense; (s) – sense

3.3 Pon3 *mRNA Expression in Wild-Type and apoE –/– Adult Mouse Hearts*

Pon3 labeling was detected in the heart wall and blood vessels. Ubiquitous labeling occurred in the wild-type mouse heart, including the heart wall (cardiac muscle), aortic root, vein, and connective tissue around the heart. This labeling was absent in the sense (control) hybridization (Fig. 11). A similar pattern of ubiquitous *Pon3* mRNA distribution was seen in the heart tissue of a mouse with the apoE –/– genotype (Fig. 12). *Pon3* mRNA was also detected in the atherosclerotic lesion areas in the aortic root of apoE –/– mouse (Fig. 12a), whereas this labeling was absent in the sense (control) hybridization (Fig. 12c). A previous report shows that mouse *Pon3* mRNA and enzyme activity is expressed by macrophages (Rosenblat et al. 2003). Therefore, the *Pon3* mRNA present in the atherosclerotic lesion area of apoE –/– mouse is probably expressed by the macrophages or foam cells present in the lesion.

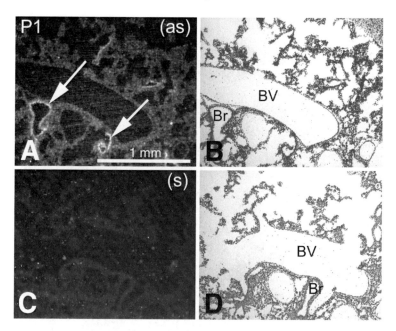

Fig. 4 *Pon3* mRNA expression in wild-type newborn (P1) mouse lung. (**a**) Emulsion autoradiography detection of *Pon3* mRNA in bronchiole epithelium (*arrows*) seen as bright labeling under brightfield illumination. (**b**) Cresyl violet staining of the same section shown in (**a**) seen under brightfield illumination. (**c**) Control hybridization in an adjacent section comparable to (**a**). (**d**) Cresyl violet staining of the same section shown in (**c**) seen under brightfield illumination. Abbreviations: Br – bronchiole; BV – blood vessel; (as) – antisense; (s) – sense

4 Discussion

Although the anti-atherogenic function of PON3 has been established using animal models (Ng et al. 2007; Shih et al. 2007), the physiological function of PON3 is still unclear. As a first step to understanding the role of PON3 in physiology, we determined the expression pattern of mouse *Pon3* mRNA in newborn and adult stages using the in situ hybridization technique. We detected high levels of mouse *Pon3* mRNA in adipose tissue, submaxillary gland, adrenal cortex, liver, lung, pancreas, stomach, and large intestine of newborn mice. *Pon3* mRNA was detected in these same tissues in adult mice, although the expression levels were lower than in the newborns. We also detected moderate levels of *Pon3* mRNA in the heart wall (cardiac muscle), aortic root, vein, and connective tissue around the heart of adult wild-type and apoE –/– mice. In apoE –/– mouse heart, *Pon3* mRNA was also present in the atherosclerotic lesion area at the aortic root. The detection of *Pon3* mRNA in the aortic root and atherosclerotic lesion areas suggests that PON3 may exert its athero-protective, anti-oxidative effect locally in the artery wall.

The developmental down-regulation of mouse *Pon3* expression between newborn and adult stages is interesting and different from what is known for PON1.

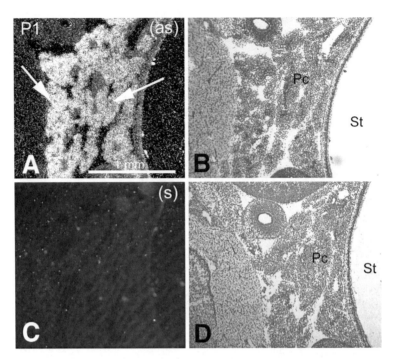

Fig. 5 *Pon3* mRNA expression in wild-type newborn (P1) mouse pancreas. (**a**) Emulsion autoradiography detection of *Pon3* mRNA in acini of the pancreatic tissue (*arrows*) seen as bright labeling under brightfield illumination. (**b**) Cresyl violet staining of the same section shown in (**a**) seen under brightfield illumination. (**c**) Control hybridization in an adjacent section comparable to (**a**). (**d**) Cresyl violet staining of the same section shown in (**c**) seen under brightfield illumination. Abbreviations: Pc – pancreas; St – stomach; (as) – antisense; (s) – sense

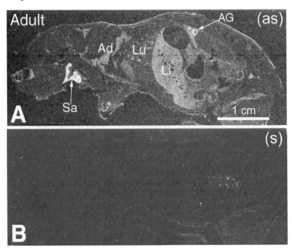

Fig. 6 *Pon3* mRNA distribution in wild-type adult mouse whole body sections. (**a**) X-ray film autoradiography detection of *Pon3* mRNA seen as bright labeling under darkfield illumination. Higher *Pon3* expression levels seem to occur in adrenal gland cortex and salivary glands (*arrows*). Less labeling is seen in adipose tissue, lung, and liver. (**b**) Control hybridization in an adjacent section comparable to (**a**). Abbreviations: Ad – adipose tissue; AG – adrenal gland; Li – liver; Lu – lung; Sa – salivary gland; (as) – antisense; (s) – sense

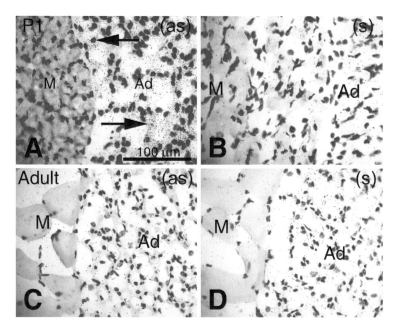

Fig. 7 Comparative *Pon3* mRNA expression in wild-type newborn (P1) and adult mouse adipose tissue. (**a**) Emulsion autoradiography detection of dense *Pon3* mRNA in newborn mouse adipose tissue (*arrows*) seen as *black precipitate* (silver grains) on a cresyl violet background under bright-field illumination. Skeletal muscle fibers are not labeled. (**b**) Control hybridization in an adjacent section comparable to (**a**). (**c**) Low-level *Pon3* mRNA labeling in the adipose tissue of an adult mouse. (**d**) Control hybridization in an adjacent section comparable to (**c**). Abbreviations: Ad – adipose tissue; M – skeletal muscle; (as) – antisense; (s) – sense

Previously we reported that human PON1 levels plateau between 6 and 15 months of age, whereas mouse PON1 levels plateau at 3 weeks of age (Cole et al. 2003). Therefore, our data suggest different developmental regulation patterns between mouse *Pon1* and *Pon3* genes, with *Pon1* being up-regulated and *Pon3* being down-regulated, respectively, as the mice reach adulthood. It remains to be seen whether the same developmental down-regulation is also true for the human *PON3* gene.

Our *Pon3* mRNA tissue distribution data are in general agreement with a recent report of the immunohistochemical analysis of PON3 expression in normal mouse tissues (Marsillach et al. 2008). Both studies found PON3 expression in mouse skin, salivary gland, glandular epithelium of stomach, intestine, hepatocytes of liver, acinar cells of pancreas, heart, adipose tissue, and bronchiole epithelium of lung. Our study showed high levels of *Pon3* mRNA present in the adrenal cortex which was not included in the immunohistochemical study (Marsillach et al. 2008). Unlike the immunohistochemical study, the *Pon3* mRNA levels in the brain, skeletal muscle, kidney, and bone were too low to be detected in our study.

Detection of *Pon3* mRNA in adipocytes suggests a possible role for PON3 in adipogenesis. In fact, our previous study showed that transgenic mice overexpressing

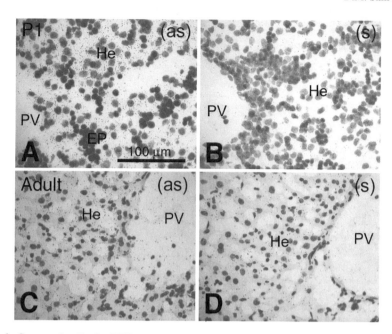

Fig. 8 Comparative *Pon3* mRNA expression in wild-type newborn (P1) and adult mouse liver. (**a**) Emulsion autoradiography detection of *Pon3* mRNA in newborn mouse liver hepatocytes seen as *black precipitate* (silver grains) on a cresyl violet background under brightfield illumination. A group of hematopoietic cells looks unlabeled. (**b**) Control hybridization in an adjacent section comparable to (**a**). (**c**) *Pon3* mRNA labeling in the liver of an adult mouse. Labeling is much less concentrated in the hepatocytes. (**d**) Control hybridization in an adjacent section comparable to (**c**). Abbreviations: EP – erythropoietic cells; He – hepatocytes; PV – portal vein; (as) – antisense; (s) – sense

human PON3 are protected against obesity (Shih et al. 2007), implicating PON3's involvement in adipocyte functions. Expression of *Pon3* in the epithelium of skin, lung, stomach, and intestine is intriguing. Since PON3 is known to hydrolyze bacterial quorum-sensing molecules, such as 3OC12-HSL (Draganov et al. 2005; Ozer et al. 2005; Yang et al. 2005), it is plausible that PON3's presence in the epithelium may play a protective role against bacterial infection. PON3's presence in the exocrine glands such as the salivary gland and the acinar cells of pancreas is very interesting given the highly secretory nature of these tissues. The secretory function of these cells relies on the capacity of the ER to fold and modify nascent polypeptides and to synthesize phospholipids for the subsequent trafficking of secretory proteins through the ER–Golgi network. Since PON3 is localized to ER (Rothem et al. 2007), we postulate that PON3 may play a role in the secretory machinery in exocrine cells.

Lastly, the high level of expression of *Pon3* mRNA in the adrenal cortex of newborn and adult mice is unexpected and intriguing. The adrenal cortex mediates the stress response through the production of mineralocorticoids and glucocorticoids,

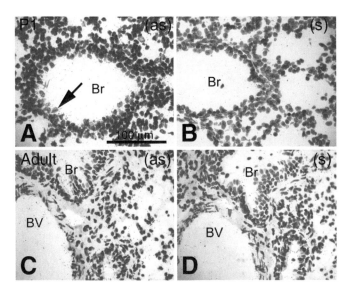

Fig. 9 Comparative *Pon3* mRNA expression in wild-type newborn (P1) and adult mouse lung. (**a**) Emulsion autoradiography detection of *Pon3* mRNA in newborn mouse bronchiole epithelium (*arrow*) seen as *black precipitate* (silver grains) on a cresyl violet background under brightfield illumination. (**b**) Control hybridization in an adjacent section comparable to (**a**). (**c**) Low level *Pon3* mRNA labeling in the bronchiole of an adult mouse. (**d**) Control hybridization in an adjacent section comparable to (**c**). Abbreviations: Br – bronchiole; BV – blood vessel; (as) – antisense; (s) – sense

including aldosterone and cortisol respectively. It is also a secondary site of androgen synthesis. Our data suggest a possible role for PON3 in steroidogenesis. We have constructed a PON3 KO mouse through gene-targeting (Stoltz et al. 2007). This mutant mouse will be useful in elucidating the function(s) of PON3 in various organs such as the adrenal gland, adipose tissue, and liver. Detailed biochemical studies are also necessary to identify the physiological substrate(s) of PON3 in various cells/tissues.

Acknowledgment We thank Yi-Shou Shi, XuPing Wang, and Phylogeny, Inc. (Columbus, OH) for excellent technical support. This work is supported by NIH grants PO1 HL30568 (to AJL and DMS), and 2RO1 HL071776-05A1 (to DMS), and AHA grant 0755069Y (to DMS).

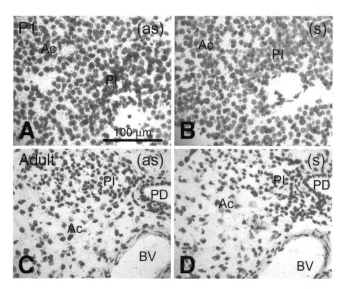

Fig. 10 Comparative *Pon3* mRNA expression in wild-type newborn (P1) and adult mouse pancreas. (**a**) Emulsion autoradiography detection of *Pon3* mRNA in newborn mouse pancreatic acinar cells seen as *black precipitate* (silver grains) on a cresyl violet background under brightfield illumination. Pancreatic islets seem to be unlabeled. (**b**) Control hybridization in an adjacent section comparable to (**a**). (**c**) Low level *Pon3* mRNA labeling in the pancreas of the adult mouse; the pancreatic islets, blood vessels, and pancreatic duct are unlabeled. (**d**) Control hybridization in an adjacent section comparable to (**c**). Abbreviations: Ac – acinar cells, BV – blood vessel; PD – pancreatic duct; PI – pancreatic islet; (as) – antisense; (s) – sense

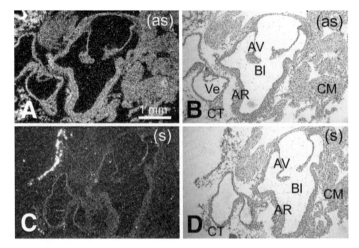

Fig. 11 *Pon3* mRNA expression in wild-type adult mouse heart. (**a**) Emulsion autoradiography detection of *Pon3* mRNA seen as bright labeling under darkfield illumination. Labeling occurs in multiple heart regions including the aortic root, aortic valve, vein, cardiac muscle, and connective tissue. (**b**) Cresyl violet staining of the same section shown in (**a**) seen under brightfield illumination. (**c**) Control hybridization in an adjacent section comparable to (**a**). (**d**) Cresyl violet staining of the same section shown in (**c**) seen under brightfield illumination. Abbreviations: AV – aortic valve, leaflet; AR – artery root; Bl – blood; CM – cardiac muscle; CT – connective tissue; En – endocardium; Ve – vein; (as) – antisense; (s) – sense

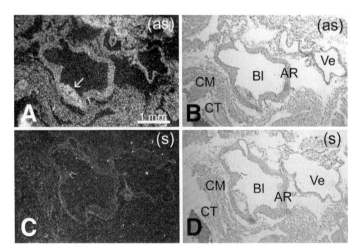

Fig. 12 *Pon3* mRNA expression in ApoE –/– adult mouse heart. (**a**) Emulsion autoradiography detection of *Pon3* mRNA seen as bright labeling under darkfield illumination. Labeling occurs in several tissues including the aortic root, vein, cardiac muscle, and connective tissue around the heart. Labeling also occurs in the atherosclerotic lesion area (*arrow*). (**b**) Cresyl violet staining of the same section shown in (**a**) seen under brightfield illumination. (**c**) Control hybridization in an adjacent section comparable to (**a**). (**d**) Cresyl violet staining of the same section shown in (**c**) seen under brightfield illumination. Abbreviations: AR – aortic root; Bl – blood; CM – cardiac muscle; CT – connective tissue; Ve – vein; (as) – antisense; (s) – sense

References

Blatter, M.C., James, R.W., Messmer, S., Barja, F., and Pometta, D. (1993). Identification of a distinct human high-density lipoprotein subspecies defined by a lipoprotein-associated protein, K-45. Identity of K-45 with paraoxonase. Eur J Biochem *211*, 871–879.

Cole, T.B., Jampsa, R.L., Walter, B.J., Arndt, T.L., Richter, R.J., Shih, D.M., Tward, A., Lusis, A.J., Jack, R.M., Costa, L.G., et al. (2003). Expression of human paraoxonase (PON1) during development. Pharmacogenetics *13*, 357–364.

Draganov, D.I., Stetson, P.L., Watson, C.E., Billecke, S.S., and La Du, B.N. (2000). Rabbit serum paraoxonase 3 (PON3) is a high density lipoprotein-associated lactonase and protects low density lipoprotein against oxidation. J Biol Chem *275*, 33435–33442.

Draganov, D.I., Teiber, J.F., Speelman, A., Osawa, Y., Sunahara, R., and La Du, B.N. (2005). Human paraoxonases (PON1, PON2, and PON3) are lactonases with overlapping and distinct substrate specificities. J Lipid Res *46*, 1239–1247.

Hassett, C., Richter, R.J., Humbert, R., Chapline, C., Crabb, J.W., Omiecinski, C.J., and Furlong, C.E. (1991). Characterization of cDNA clones encoding rabbit and human serum paraoxonase: the mature protein retains its signal sequence. Biochemistry *30*, 10141–10149.

Marsillach, J., Mackness, B., Mackness, M., Riu, F., Beltran, R., Joven, J., and Camps, J. (2008). Immunohistochemical analysis of paraoxonases-1, 2, and 3 expression in normal mouse tissues. Free Radic Biol Med *45*, 146–157.

Ng, C.J., Bourquard, N., Hama, S.Y., Shih, D., Grijalva, V.R., Navab, M., Fogelman, A.M., and Reddy, S.T. (2007). Adenovirus-mediated expression of human paraoxonase 3 protects against the progression of atherosclerosis in apolipoprotein E-deficient mice. Arterioscler Thromb Vasc Biol *27*, 1368–1374.

Ng, C.J., Wadleigh, D.J., Gangopadhyay, A., Hama, S., Grijalva, V.R., Navab, M., Fogelman, A.M., and Reddy, S.T. (2001). Paraoxonase-2 is a ubiquitously expressed protein with antioxidant properties and is capable of preventing cell-mediated oxidative modification of low density lipoprotein. J Biol Chem *276*, 44444–44449.

Ozer, E.A., Pezzulo, A., Shih, D.M., Chun, C., Furlong, C., Lusis, A.J., Greenberg, E.P., and Zabner, J. (2005). Human and murine paraoxonase 1 are host modulators of Pseudomonas aeruginosa quorum-sensing. FEMS Microbiol Lett *253*, 29–37.

Parsek, M.R., and Greenberg, E.P. (2000). Acyl-homoserine lactone quorum sensing in gram-negative bacteria: a signaling mechanism involved in associations with higher organisms. Proc Natl Acad Sci U S A *97*, 8789–8793.

Reddy, S.T., Wadleigh, D.J., Grijalva, V., Ng, C., Hama, S., Gangopadhyay, A., Shih, D.M., Lusis, A.J., Navab, M., and Fogelman, A.M. (2001). Human paraoxonase-3 is an HDL-associated enzyme with biological activity similar to paraoxonase-1 protein but is not regulated by oxidized lipids. Arterioscler Thromb Vasc Biol *21*, 542–547.

Rosenblat, M., Draganov, D., Watson, C.E., Bisgaier, C.L., La Du, B.N., and Aviram, M. (2003). Mouse macrophage paraoxonase 2 activity is increased whereas cellular paraoxonase 3 activity is decreased under oxidative stress. Arterioscler Thromb Vasc Biol *23*, 468–474.

Rothem, L., Hartman, C., Dahan, A., Lachter, J., Eliakim, R., and Shamir, R. (2007). Paraoxonases are associated with intestinal inflammatory diseases and intracellularly localized to the endoplasmic reticulum. Free Radic Biol Med *43*, 730–739.

Shamir, R., Hartman, C., Karry, R., Pavlotzky, E., Eliakim, R., Lachter, J., Suissa, A., and Aviram, M. (2005). Paraoxonases (PONs) 1, 2, and 3 are expressed in human and mouse gastrointestinal tract and in Caco-2 cell line: selective secretion of PON1 and PON2. Free Radic Biol Med *39*, 336–344.

Shih, D.M., Xia, Y.R., Wang, X.P., Wang, S.S., Bourquard, N., Fogelman, A.M., Lusis, A.J., and Reddy, S.T. (2007). Decreased obesity and atherosclerosis in human paraoxonase 3 transgenic mice. Circ Res *100*, 1200–1207.

Stoltz, D.A., Ozer, E.A., Ng, C.J., Yu, J.M., Reddy, S.T., Lusis, A.J., Bourquard, N., Parsek, M.R., Zabner, J., and Shih, D.M. (2007). Paraoxonase-2 deficiency enhances Pseudomonas aeruginosa quorum sensing in murine tracheal epithelia. Am J Physiol Lung Cell Mol Physiol *292*, L852–L860.

Teiber, J.F., Billecke, S.S., La Du, B.N., and Draganov, D.I. (2007). Estrogen esters as substrates for human paraoxonases. Arch Biochem Biophys *461*, 24–29.

Yang, F., Wang, L.H., Wang, J., Dong, Y.H., Hu, J.Y., and Zhang, L.H. (2005). Quorum quenching enzyme activity is widely conserved in the sera of mammalian species. FEBS Lett *579*, 3713–3717.

PON1 and Oxidative Stress in Human Sepsis and an Animal Model of Sepsis

Dragomir Draganov, John Teiber, Catherine Watson, Charles Bisgaier, Jean Nemzek, Daniel Remick, Theodore Standiford, and Bert La Du

Abstract Sepsis is the leading cause of death in critically ill patients. The pathophysiological mechanisms implicated in the development of sepsis and organ failure are complex and involve activation of systemic inflammatory response and coagulation together with endothelial dysfunction. Oxidative stress is a major promoter and mediator of the systemic inflammatory response. Serum PON1 has been demonstrated in multiple clinical and animal studies to protect against oxidative stress, but also to undergo inactivation upon that condition. We found decreased plasma PON1 activity in patients with sepsis compared to healthy controls or critically ill patients without sepsis; furthermore, in sepsis patients PON1 activity was lower and remained lower in the course of sepsis in the non-survivors compared to the survivors. Plasma PON1 activity was positively correlated with high-density lipoprotein cholesterol and negatively correlated with markers of lipid peroxidation. In an experimental animal model of sepsis, murine cecal ligation and puncture, the time course of plasma PON1 activity was very similar to that found in sepsis patients. Persistently low PON1 activity in plasma was associated with lethal outcome in human and murine sepsis.

Keywords Sepsis · Systemic inflammatory response · Cecal ligation and puncture model · Critically ill patients · Oxidative stress · High-density lipoprotein · Lipid peroxides · Total antioxidant activity · Thiobarbituric acid reacting substances

Abbreviations

ArE	arylesterase activity
G	(needle size) gauge
HDL	high-density lipoprotein
ICU	Intensive Care Unit
IL-6	interleukin-6

D. Draganov (✉)
Head of Toxicokinetics, Metabolism Department, WIL Research Laboratories, LLC, Ashland, OH, 44805, USA
e-mail: ddraganov@wilresearch.com

S.T. Reddy (ed.), *Paraoxonases in Inflammation, Infection, and Toxicology*, Advances in Experimental Medicine and Biology 660, DOI 10.1007/978-1-60761-350-3_9,
© Humana Press, a part of Springer Science+Business Media, LLC 2010

LPO   lipid peroxides
RNS   reactive nitrogen species
ROS   reactive oxygen species
SIRS   systemic inflammatory response
TAA   total antioxidant activity
TBARS   thiobarbituric acid reacting substances

1 Introduction

Sepsis is the leading cause of death in critically ill patients (Martin et al. 2003). The pathophysiological mechanisms implicated in the development of sepsis and organ failure are complex. The recognition of the pathogen and its byproducts by Toll-like receptors activates the release of various pro-inflammatory cytokines and mediators which then activate the endothelium, white blood cells and epithelial cells causing the production and release of vasoactive mediators such as thromboxane, prostaglandins, nitric oxide and free oxygen radicals (De Backer, 2007). Although these mediators play an important role in the defense against infection, the uncontrolled activation of the anti-inflammatory response may lead to profound hemodynamic and metabolic effects, which in turn may lead to organ dysfunction and failure (Abraham and Singer, 2007). Oxidative stress, defined as generation of reactive nitrogen or oxygen species (RNS and ROS, respectively) in excess of the antioxidant capacity of the body, is a major promoter and mediator of the systemic inflammatory response (SIRS). Sustained and excessive production of ROS and RNS is associated with increased morbidity and mortality in sepsis patients (Goode et al. 1995; Berger and Chiolero, 2007).

Serum PON1 is a high-density lipoprotein (HDL)-associated enzyme that has been demonstrated in multiple clinical and animal studies to protect against oxidative stress, whereby PON1 undergoes inactivation (Aviram and Rosenblat, 2004; Aviram and Rosenblat, 2005; Ng et al. 2005). The studies described in this chapter are aimed at evaluation of plasma PON1 activity in patients with sepsis and in an experimental animal model of sepsis, and its relation to markers of oxidative stress and outcome in sepsis.

2 PON1 and Lipid Peroxidation in Sepsis Patients and Healthy Controls

In a pilot study, blood samples were obtained from six consecutive sepsis patients from an Intensive Care Unit (ICU) at the University of Michigan Hospital and six age- and sex-matched outpatients who were apparently healthy. Plasma was isolated by centrifugation and assayed for arylesterase activity (ArE), lipid and lipoprotein metabolism, lipid peroxides (LPO) and thiobarbituric acid reacting substances (TBARS). Compared to controls, septic patients had increased triglycerides, LPO

Table 1 Plasma PON1, lipids and lipid peroxidation in sepsis patients and healthy controls

Parameter	Units	Controls ($n = 6$) (mean ± SD)	Sepsis Patients ($n = 6$) (mean ± SD)	Correlation with ArE (r)
PON1 (ArE)	U/mL	107 ± 36.1	39.6 ± 13.7*	NA
HDL Cholesterol	mg/dL	45.0 ± 14.5	13.3 ± 8.24*	0.89**
Triglycerides	mg/dL	125 ± 50.4	225 ± 112	−0.50
LPO	nmol/mL	9.96 ± 8.68	42.0 ± 31.2*	−0.70**
TBARS	nmol/mL	1.97 ± 0.626	5.83 ± 3.30	−0.69**

*Significantly different from controls at $p < 0.05$.
**Significant at $p < 0.05$.
NA = Not applicable.

Fig. 1 Plasma PON1 activity in sepsis patients and age-matched healthy controls. *Bars* represent median values, *error bars* represent range. ArE activity with phenyl acetate, 1 U = 1 μmol/min

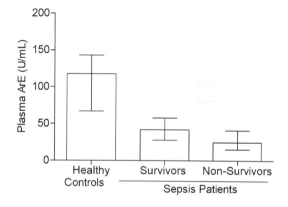

and TBARS and significantly decreased HDL cholesterol levels and ArE (Table 1). ArE was positively correlated with HDL cholesterol and negatively with LPO and TBARS ($p < 0.05$). When more patients and controls were enrolled in the study, PON1 ArE was confirmed significantly lower in sepsis patients (median 37 U/ml, range 15–63 U/ml, $n = 15$) than in controls (median 118 U/ml, range 67–143 U/ml, $n = 20$) (Fig. 1). Furthermore, plasma PON1 activity was even lower in the non-survivors, and we observed that all patients with plasma ArE below 35 U/ml died.

3 PON1 and LPO in ICU Patients with and Without Sepsis

To follow up on the results from the pilot study, we designed a prospective observational study to further explore the relationship between PON1 enzyme levels and the sepsis progression. One of the objectives of the study was to test the hypothesis that PON levels and activities are inversely correlated with evidence of enhanced systemic inflammation and oxidative stress. Patients ≥18 years old of both sexes were recruited at the University of Michigan Medical ICU at the onset of sepsis. Sepsis was defined by the presence of at least two signs of SIRS (temperature $\geq 38°C$ or $\leq 36°C$; heart rate ≥ 90 beats per minute; respiratory rate ≥ 20 per minute or $PaCO_2$ ≤ 32 mmHg; white blood cells $\geq 12,000/mm^3$ or $\leq 4,000/mm^3$ or $\geq 10\%$ immature neutrophils). Patients were excluded if they were pregnant, had significant liver disease or HIV infection or had recently received corticosteroid, cytotoxic or investigational drug therapy. For the non-septic control group we enrolled patients ≥ 18 years old of both sexes admitted to the ICU for disorders other than sepsis and who did not have any of the exclusion criteria outlined above. Prior to entry into the study, an informed consent was properly obtained from each patient or patient's legally acceptable representative. The study was approved by the Institutional Medical Board Review Committee. At the time of entry, a complete medical/sepsis history and physical examination was obtained from each subject. Blood samples were drawn at study entry (day 0) and 1, 3, 7 and 14 days post-entry for determination of, among other parameters, PON1 activity, lipid and lipoprotein metabolism, LPO and TBARS. A total of 124 ICU patients, of whom 85 were with sepsis and 39 without sepsis, were recruited in the study. The demographic characteristics and other results from this study are to be reported elsewhere (Draganov et al. 2008, manuscript in preparation); here we present plasma PON1 ArE, lipids and lipid peroxidation data (Table 2). Compared to ICU patients without sepsis, the sepsis patients had lower ArE and HDL cholesterol and higher triglycerides, LPO and TBARS at study entry and throughout the course of observation. Among the sepsis patients, non-survivors had lower ArE and HDL cholesterol and higher triglycerides, LPO and TBARS than survivors; however, these differences were not significant at study entry ($p > 0.05$). ArE activity was positively correlated with HDL cholesterol in both sepsis and non-sepsis patients and with triglycerides only in the non-sepsis patients; no other significant correlations with the parameters presented in Table 2 were found. In the time course of observation, plasma ArE and HDL cholesterol remained

Table 2 Plasma PON1, lipids and lipid peroxidation in ICU patients with and without sepsis

Parameter	Units	No sepsis	Sepsis		
			All	Survivors	Non-survivors
PON1 (ArE)	U/mL	65.7 (63.3)	48.7* (46.5)*	50.1 (48.2)	44.3 (39.3)**
HDL Cholesterol	mg/dL	36.1 (36.4)	21.7* (21.1)*	22.6 (21.9)	19.9 (17.9)
Triglycerides	mg/dL	127 (131)	159 (169)*	156 (162)	171 (203)
LPO	nmol/mL	83.1 (76.7)	176* (167)*	169 (153)	209 (229)**
TBARS	nmol/mL	1.14 (1.15)	1.33 (1.31)*	1.30 (1.29)	1.47 (1.51)**

Presented values are at study entry or for the whole data set (reported in brackets); the number of samples analyzed per group were as follows: non-septic patients 38 (76), sepsis patients 85 (215), survivors 69 (180), and non-survivors 16 (35).
*Significantly different from non-sepsis patients at $p < 0.05$.
** Significantly different from survivors at $p < 0.05$.
NA = Not applicable.

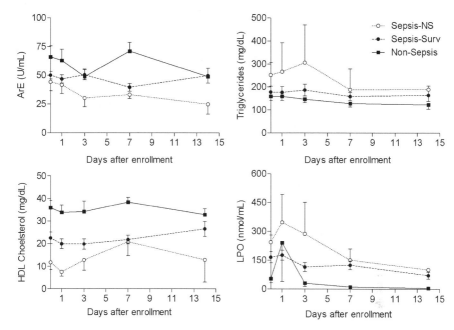

Fig. 2 Time course of plasma PON1 activity, lipids and LPO in ICU patients with and without sepsis

lower in the sepsis patients compared to the critically ill patients without sepsis (Fig. 2). Among the sepsis patients, the PON1 and HDL cholesterol levels were lower in the non-survivors than in the survivors. The reverse trend was observed in the time course of the triglycerides and LPO levels – they were higher in the sepsis versus non-sepsis patients and sepsis-non-survivors versus sepsis-survivors (Fig. 2).

4 PON1 and LPO in Murine Cecal Ligation and Puncture Model of Sepsis

Patients with sepsis are represented by a heterogeneous population that differs by age, gender, genetic factors, and by an array of accompanying disorders and diseases. In addition, there is a considerable diversity in pathogens infecting these patients, their virulence and antimicrobial sensitivity. Therefore, to better characterize the mechanisms underlying the regulation of the inflammatory response and to develop successful treatments for sepsis and septic shock, animal models of sepsis are used, in which standardized and controlled conditions can be employed. In the murine cecal ligation and puncture (CLP) model of sepsis, the slow leakage of intestinal flora in the abdominal cavity produces gradual disease development that closely mimics human sepsis (Remick et al. 2000). The severity and progression of the sepsis in this model is proportional to the size of the needle used (bigger gauge number indicates lower needle size). Thus, puncture with a 25-gauge (G) needle leads to mild or absent systemic inflammation and less than 10% of the animals develop sepsis; puncture with a 21-G needle induces sepsis in about 90% of the animals with approximately 50% lethality; and puncture with a 18-G needle is lethal with all treated animals developing sepsis and dying within 2–3 days after the operation (Remick et al. 2002). Female wild-type BALB/c mice were used in all the experiments described in this chapter; all operated animals received standard antibiotic and fluid resuscitation treatment, which reflects the current treatment of sepsis patients at ICU. In one set of experiments, PON1 activity and oxidative stress parameters in plasma and tissues were assessed in mild-inflammation (25-G CLP) and lethal SIRS (18-G CLP). Animals from both groups were sacrificed at 6, 16 and 24 h after CLP operation (three mice per time point) and blood samples and selected tissues were collected. All animals from the 25-G group appeared to recover within 4–6 h after the CLP and their behavior was similar to that of control (non-operated) animals. In contrast, the 18-G group showed progressive impairment, the animals remained lethargic with respiratory difficulties and showed typical SIRS changes such as increased number of lymphocytes and decreased neutrophils and platelets. Interleukin-6 (IL-6) levels were significantly increased only in the 18-G group, ranging from 1,000 to 32,000 pg/mL at the different time points. HDL cholesterol decreased approximately 40% in both the 25-G and 18-G groups at 6 h after CLP, then increased to 150–180% of the basal level in the 25-G group, but remained below basal levels in the 18-G group (Fig. 3). LPO and TBARS in plasma were slightly elevated at all time points for both groups (data not shown). Total antioxidant activity (TAA) in plasma decreased progressively in the 18-G group, but not the 25-G group (Fig. 3). PON1 activities with phenyl acetate (ArE) and paraoxon were highly correlated in plasma and tissue homogenates; only paraoxonase activity data are presented here. Plasma paraoxonase activity decreased by 30% at 6 h post-CLP in both groups, then restored to 80% of the control at the later time points in the 25-G group, but decreased further in the 18 G group (Fig. 3). In the liver homogenates, paraoxonase activity in the 25-G group showed a reverse trend to the plasma levels, which together with the increased HDL cholesterol levels in plasma is suggestive of an increased secretion (liver is the main source of PON1 in circulation). In the 18-G group, liver paraoxonase activity remained relatively steady throughout the course of observation. Thus, the decreased plasma paraxonase activity in the 18-G group was likely due to inactivation of the enzyme which was paralleled by the decreased plasma TAA.

In another set of experiments, plasma PON1 activity (with paraoxon) was followed in the time course of sepsis development (21-G CLP) (Fig. 4). Plasma paraoxonase activity decreased to less than 50% of the basal levels at day 2 after CLP. In the survivors, paraoxonase activity recovered to about 80% of the basal level by day 7, but failed to recover in the non-survivors. The changes in the plasma PON1 activity in the septic mice were similar to those we have observed in sepsis patients.

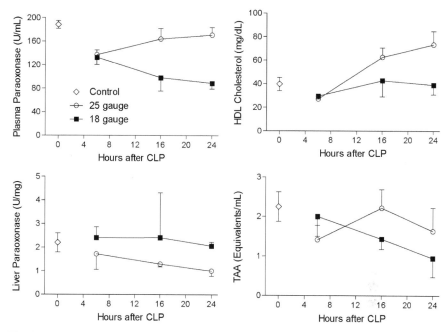

Fig. 3 Plasma and liver PON1 activity, plasma HDL cholesterol and TAA in BALB/c mice after Lethal (18-G) and non-lethal (25-G) CLP. Paraoxonase activity with paraoxon, 1 U = 1 nmol/min. TAA activity is expressed as Trolox equivalents. Each time point represents mean ± standard deviation ($n = 3$).

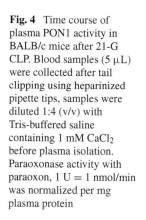
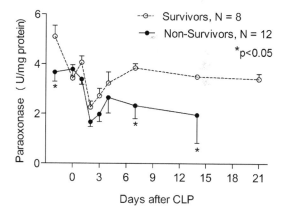

Fig. 4 Time course of plasma PON1 activity in BALB/c mice after 21-G CLP. Blood samples (5 μL) were collected after tail clipping using heparinized pipette tips; samples were diluted 1:4 (v/v) with Tris-buffered saline containing 1 mM CaCl$_2$ before plasma isolation. Paraoxonase activity with paraoxon, 1 U = 1 nmol/min was normalized per mg plasma protein

5 Conclusions

Profound changes in the concentration and composition of plasma lipids and lipoproteins have been described in acute infection, inflammatory disease, critical illness and in ICU patients with sepsis and septic shock (Van Leeuwen et al. 2003;

Wu et al. 2004). Lipid and lipoprotein concentrations were reduced (by 30% or more) in critically ill surgical patients (Gordon et al. 2001). These changes are the result of the release of pro-inflammatory cytokines during the acute response; in sepsis, these cytokines also induce increased hepatic synthesis of triglyceride-rich lipoproteins dubbed "lipemia of sepsis" (Wu et al. 2004). In patients with severe sepsis, lipoprotein concentration changed rapidly and was reduced to up to 50% of recovery concentrations (Gordon et al. 1996). Low HDL cholesterol on day 1 of severe sepsis was found to be significantly associated with increase in mortality and adverse clinical outcomes (Chien et al. 2005). Our finding from the studies in sepsis patients versus healthy controls and in ICU patients with and without sepsis are concordant with the literature. HDL cholesterol levels were significantly lower in sepsis patients compared to either healthy controls or critically ill patients without sepsis. PON1 ArE was strongly correlated with HDL cholesterol in both studies and displayed the same trend; furthermore, ArE was significantly lower in non-survivors compared to survivors among the sepsis patients (Table 2). HDL is important for PON1 secretion and stability and stimulates PON1's enzymatic activities (James and Deakin, 2004). On the other hand, PON1 is an important component of HDL, conferring HDL's anti-inflammatory and anti-oxidant properties (Ng et al. 2005). During acute phase response, however, alterations in HDL protein composition and oxidative modifications of its major protein, apolipoprotein A-1, convert HDL to pro-inflammatory particles (Navab et al. 2007). Thus, PON1 activity appears to be a better determinant of outcome in human sepsis than HDL cholesterol. This speculation is supported by our findings in the murine CLP model of sepsis, where the temporal changes in the plasma paraoxonase activity but not in HDL cholesterol corresponded to the impairment of the animals in the first 24 h after a lethal challenge (18-G CLP). In sepsis patients, PON1 levels before the onset of sepsis are not known, therefore it is not clear if pre-sepsis PON1 activity was lower and/or there was a bigger drop in PON1 activity in the course of sepsis in the non-survivors compared to the survivors. Thus, it was intriguing to observe significantly lower basal PON1 activity in the mice which died after 21-G CLP compared to the survivors (Fig. 4). There are multiple genetic, environmental, diet and disease state-related factors affecting plasma PON1 activity (Costa et al. 2005). In the human studies, we found that plasma PON1 activity was inversely correlated with LPO and TBARS; in mouse 18-G CLP, plasma PON1 activity was positively correlated with TAA. The cause–effect relationship between decreased PON1 activity and increased oxidative stress is not established; most likely it is a dynamic bi-directional relationship.

In conclusion, persistently low PON1 activity in plasma was associated with lethal outcome in human and murine sepsis. The usefulness and feasibility of plasma PON1 activity as a trigger for more aggressive therapy and/or as a measure for success of a therapeutic regimen in ICU patients will need further evaluation.

Acknowledgments The contribution of Scott Billecke, Audrey Speelman, Gerry Bolgos, Dennis Cooperson, Michael Newsteadt (University of Michigan), Kristin Sass and Sandra Drake (Esperion Therapeutics) is gratefully acknowledged. This work was supported by Michigan Life Sciences Corridor Fund No. 001796.

References

Abraham E, Singer M (2007) Mechanisms of sepsis-induced organ dysfunction. Crit Care Med 35:2408–2416

Aviram M, Rosenblat M. (2004) Paraoxonases 1, 2, and 3, oxidative stress, and macrophage foam cell formation during atherosclerosis development. Free Radic Biol Med 37:1304–1316

Aviram M, Rosenblat M. (2005) Paraoxonases and cardiovascular diseases: pharmacological and nutritional influences. Curr Opin Lipidol 16:393–399

Berger M, Chiolero RL (2007) Antioxidant supplementation in sepsis and systemic inflammatory response syndrome. Crit Care Med 35(Suppl):584–590

Chien J-Y, Jerng J-S, Yu C-J et al. (2005) Low serum level of high-density lipoprotein cholesterol is a poor prognostic factor for severe sepsis. Crit Care Med 33:1688–1693

Costa LG, Vitalone A, Cole TB et al. (2005) Modulation of paraoxonase (PON1) activity. Biochem Pharmacol 69:541–550

De Backer D (2007) Benefit-risk assessment of drotrecogin alfa (activated) in the treatment of sepsis. Drug Saf 30:995–1010

Draganov DI, Teiber JF, Billecke SS et al. (2008) Serum paraoxonase (PON1) and apolipoprotein A-I (apoA-I) are reduced in critically ill patients with sepsis. Manuscript in preparation.

Goode HF, Cowley HC, Walker BE et al. (1995) Decreased antioxidant status and increased lipid peroxidation in patients with septic shock and secondary organ dysfunction. Crit Care Med 23:646–651

Gordon BR, Parker TS, Levine DM et al. (1996) Low lipid concentrations in critically illness: implications for preventing and treating endotoximia. Crit Care Med 24:584–589

Gordon BR, Parker TS, Levine DM et al. (2001) Relationship of hypolipidemia in cytokine concentrations and outcome in critically ill surgical patients. Crit Care Med 29:1563–1568

Martin GS, Mannino DM, Eaton S et al. (2003) The epidemiology of sepsis in the United States from 1979 through 2000. N Engl J Med 348:1546–1554

James RW, Deakin SP (2004) The importance of high-density lipoproteins for paraoxonase-1 secretion, stability, and activity. Free Radic Biol Med 15:1986–1994

Navab M, Yu R, Gharavi N, Huang W et al. (2007) High-density lipoprotein: antioxidant and anti-inflammatory properties. Curr Atheroscler Rep 9:244–248

Ng CJ, Shih DM, Hama SY et al. (2005) The paraoxonase gene family and atherosclerosis. Free Radic Biol Med 38:153–163

Remick DG, Newcomb DE, Bolgos GL et al. (2000) Comparison of the mortality and inflammatory response of two models of sepsis: lipopolysacharide vs. cecal ligation and puncture. Shock 13:110–116

Remick DG, Bolgos GR, Siddiqui J et al. (2002) Six at six: interleukin-6 measured 6 h after the initiation of sepsis predicts mortality over 3 days. Shock 17:463–467

Van Leeuwen HJ, Heezius EC, Dallinga GM et al. (2003) Lipoprotein metabolism in patients with severe sepsis. Crit Care Med 31:1359–1366

Wu A, Hinds CJ, Thiemermann C (2004) High-density lipoproteins in sepsis and septic shock: metabolism, actions, and therapeutic applications. Shock 21:210–221

Paraoxonase 1 Attenuates Human Plaque Atherogenicity: Relevance to the Enzyme Lactonase Activity

Hagai Tavori, Jacob Vaya, and Michael Aviram

Abstract Human atherosclerotic lesions contain a variety of lipids and oxidized lipids, which can induce atherogenic properties such as macrophage oxidation, lipoprotein oxidation and inhibition of cholesterol efflux from macrophages. These atherogenic properties of the plaque's lipid fraction are associated with the inhibition of paraoxonase 1 (PON1) lactonase activity. In contrast, incubation of PON1 with the plaque's lipid fraction reduces the lesion's atherogenic properties by lowering the capacity of the oxidized lipids to induce further oxidation. The mechanism of PON1's protective action and its endogenous substrate however remain elusive. Modeling studies may characterize PON1's possible active site, and help envisage the structure of potential endogenous and exogenous lactones as PON1 ligands. Such modeling thus may lead to a better understanding of PON1's anti-atherogenic mechanism of action.

Keywords Atherosclerosis · Docking · Lactones · Modeling · Oxidative stress · Paraoxonase · Plaque

Abbreviations

CVD	cardiovascular disease
EtAc	ethyl acetate
HDL	high density lipoprotein
7-keto-ch	7-ketocholesterol
LDL	low density lipoproteins
LE	lesion extract
LT	*N*-linoleoyl tyrosine
LTG	*N*-linoleoyl tyrosine 2'-deoxyguanosyl ester
MPM	mouse peritoneal macrophages
7-OH-ch	7-hydroxycholesterols

H. Tavori (✉)
The Lipid Research Laboratory, Technion Faculty of Medicine, Rappaport Family Institute for Research in the Medical Science and Rambam Medical Center, 31096, Haifa, Israel
e-mail: hagait@migal.org.il

S.T. Reddy (ed.), *Paraoxonases in Inflammation, Infection, and Toxicology*, Advances in Experimental Medicine and Biology 660, DOI 10.1007/978-1-60761-350-3_10,
© Humana Press, a part of Springer Science+Business Media, LLC 2010

7-OOH-ch   7-hydroperoxycholesterols
OS   oxidative-stress
Ox-LDL   oxidized low density lipoproteins
PON1   paraoxonase 1
rePON1   recombinant PON1
TBARS   thiobarbituric acid-reactive substance

1 Atherosclerotic Plaque and Oxidative Stress

Atherosclerotic plaque formation and progression is associated with endothelial cell dysfunction, accumulation of lipoprotein aggregates in the intima, followed by monocyte migration through the endothelium and their differentiation into macrophages (Lusis, 2000). The progression of the atherosclerotic plaque is characterized by increasing levels of Ox-lipoprotein, phospholipids, triglyceride (Tg) and by accumulation of proteins such as fibrinogen, apo-AI, clusterin and paraoxonase 1 (PON1) in the arterial wall, along with the progression of the disease from fatty streak to advanced atherosclerotic plaque (Mackness et al. 1997). Reactive oxygen species (ROS) and reactive nitrogen species (RNS) are both linked to the development of atherosclerosis, due to increased production of oxidants, and to a decreased total antioxidant capacity (Paravicini and Touyz, 2008). The identification of oxidative stress (OS) in an organ is important for the characterization of OS and for early intervention to delay or stop the development of pathological conditions. For this end, many endogenous markers are in use and others are under intensive investigation. We have previously developed specific exogenous markers, designed to trap ROS/RNS and to form oxidized probes, which can be used as a fingerprint characterizing the OS (Fig. 1a, b) (Khatib and Musa, 2007; Szuchman et al. 2008; Szuchman et al. 2006).

1.1 Atherosclerotic Plaque Composition

Human atherosclerotic plaque contains oxidized fatty acids, oxidized and aggregated low density lipoproteins (LDL) (Khan-Merchant et al. 2002), oxidized cholesterol products (oxysterols) (Vaya et al. 2001), lipid peroxides (Aviram et al. 2000), proteins, triglycerides (Mackness et al. 1997; Stadler et al. 2008) and antioxidant decreased levels (Aviram et al. 1995). Oxidized fatty acids promote atherosclerosis development, due to both elevation in OS and changes in plasma cholesterol profile (Khan-Merchant et al. 2002). Oxidized LDL (Ox-LDL) promote plaque formation by various means, including recruitment of monocytes to the vessel wall, increased uptake of atherogenic lipoprotein by macrophages, and enhanced foam cell formation (Stocker and Keaney, 2004). Other oxidized molecules such as oxysterols and lipid peroxides have numerous effects which are associated with atherosclerosis, such as regulation of cell death, cell growth, activation of certain kinases, and

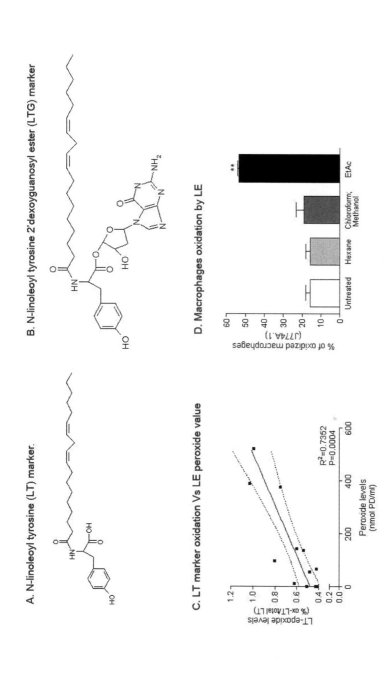

Fig. 1 Oxidative stress measurements. (**a**) Structure of *N*-linoleoyl tyrosine (LT) exogenous OS marker. (**b**) Structure of *N*-linoleoyl tyrosine 2'-deoxyguanosyl ester (LTG) exogenous OS marker. (**c**) Lesion extract incubated with the LT probe overnight; levels of LT-epoxide plotted versus peroxide concentration. Line shows direct correlation with $R^2 = 0.7352$ ($p = 0.0004$). (**d**) Oxidative status of macrophages after 24 h incubation with fractions of carotid plaques extracted with different solvents. Results expressed as % oxidized cell. One-way ANOVA analysis was performed for the above statistical analyses. Significant differences from untreated macrophages **$p < 0.001$

expression of pro-inflammatory proteins (Guardiola et al. 1996; Skoczynska, 2005; Vaya et al. 2001). It is believed that plaque formation shares a common mechanism of initial oxidative steps that leads to the generation of early fatty streak plaques (Parthasarathy et al. 1781), yet the composition of atherosclerotic plaques from different individuals can vary extensively. At the cellular level, high-risk lesions ("vulnerable plaque") have a higher content of macrophages and inflammatory cells (Ibanez et al. 2007) and different plaques were shown to contains different levels of lipids and of oxidized lipids (Tavori et al. 2008). In that respect, we observed that differences in plaque peroxidized lipids correlates with marked changes in the plaque's ability to induce oxidation, as measured by our specific exogenous oxidation probes (Fig. 1c).

1.2 Human Carotid Plaque Atherogenicity

We have shown that the human carotid plaque lipid fraction (ethyl-acetate extract) contains large amount of lipid peroxides. Following incubation of this lesion lipid extract (LE) with LDL, increase in the thiobarbituric acid reactive substance (TBARS) levels of LDL in a dose response manner was observed (Tavori et al. 2009). Oxidative modification of LDL is a well-known key event during early atherogenesis. Our results further indicate that incubation of a specific LE with macrophages induced their subsequent oxidation, whereas other lipid fractions (chloroform, methanol and hexane) of the atherosclerotic plaque did not significantly change the macrophages' oxidative status (Fig. 1c). Macrophage oxidation is known to be associated with increased levels of intracellular lipid peroxides and elevated cell-mediated oxidation of LDL, which enhance the release of cytokines such as interleukin 1, and inhibit the release of apolipoprotein E from the cells (Fuhrman et al. 1997; Fuhrman et al. 1994; Glass and Witztum, 2001). In terms of biological activities, we were able to demonstrate that the LE is indeed able to inhibit HDL ability to stimulate efflux of cholesterol from macrophages, an essential event in cholesterol homeostasis. This inhibition of HDL-mediated cholesterol efflux correlates with the inhibition of PON1 lactonase activity ($R^2 = 0.8929$, $p = 0.0086$). Thus, we may claim that the atherosclerotic plaque with its oxidized lipids serves as an atherogenic agent.

2 Paraoxonase 1 Protection Against Atherosclerosis

Paraoxonase 1 is a calcium-dependent HDL-bound enzyme, synthesized mostly in the liver and secreted into the blood stream. This enzyme catalyzes the hydrolysis of multiple compounds, such as organophosphates, arylesters, and lactones (Billecke, 2000; Draganov and La Du, 2004; Jakubowski et al. 2000; La Du, 1992). Many of the anti-atherogenic functions of HDL are attributed to PON1 (Aviram, 2000; Mackness et al. 2006; Rosenblat et al. 2006]; Rozenberg et al. 2003). PON1 can specifically bind to macrophages (Efrat and Aviram, 2008), enhances cholesterol efflux from macrophages as a function of its concentration (Rosenblat et al. 2005), protects LDL oxidation by lowering lipid peroxide levels, inhibits Ox-LDL uptake by macrophages (thus inhibiting macrophage foam cell formation), and inhibits macrophage cholesterol biosynthesis (Rozenberg et al. 2003). PON1 was also shown to diminish levels of specific oxidized lipids (Navab et al. 1996; Watson et al. 1995) and cholesterol ester hydroperoxide (Aviram et al. 2000) in lesions, macropages, and lipoproteins. Additionally PON1 was suggested to hydrolyze phosphatidylcholine isoprostanes and core aldehydes (Ahmed et al. 2001; Ahmed et al. 2002), although some of these activities were later shown to belong to platelet-activating factor acetylhydrolase (PAF-AH) (Kriska et al. 2007; Marathe et al. 2003). PON1 levels and activity have been shown to be regulated genetically, nutritionally and pharmacologically (Aviram et al. 2005; Costa et al. 2005; Deakin and James, 2004; Efrat et al. 2008). Decreased levels of PON1 were observed in hypercholesterolemic, diabetic and cardiovascular disease (CVD) patients (Rosenblat et al. 2006). PON1-deficient mice are susceptible to the development of atherosclerosis (Rozenberg et al. 2003; Shih et al. 1995), which is associated with increased macrophage OS. Blocking PON1 binding to HDL by specific antibody significantly reduces the anti-atherogenic properties of the enzyme (Efrat and Aviram, 2008). In contrast, over-expression of human PON1 in transgenic mice inhibited the development of atherosclerosis (Mackness et al. 2006). Correspondingly, the process of plaque regression has been reported to be associated with lesion destruction within the arterial wall (Ibanez et al. 2007) by a mechanism involving lower levels and retention of apo-B100 lipoproteins in the arterial wall, an enhanced rate of cholesterol efflux, and migration of foam cells out of the arterial wall (Williams et al. 2008). We thus assume that agents that would reduce the levels of lipid peroxides, such as PON1, might also reduce human carotid plaque atherogenicity.

2.1 Paraoxonase 1 in Human Atherosclerotic Plaque

PON1 was shown to accumulate in the arterial wall as human plaque progresses from fatty streak into advanced lesion (Mackness et al. 1997). PON1 accumulation in lesions was recently suggested to occur in macrophages. The enzyme accumulation in plaques was suggested to provide a protective mechanism against increasing oxidation associated with plaque progression (Mackness et al. 1997). We observed that pretreatment of human carotid lesion extract (LE), with recombinant PON1 (rePON1) resulted in a significantly reduced LE oxidative potential. Furthermore, PON1 decreased plaque atherogenicity, as indicated by the lower LE capacity to oxidatively modify macrophages (MPMs and the J774A.1 cell line) (Tavori et al. 2008). Incubation of HDL with LE that had been pretreated with rePON1 for 24 h also improved the HDL ability to stimulate cholesterol efflux from macrophages. Our findings may explain, at least in part, the well documented anti-atherogenic properties of PON1, such as the protection of LDL and HDL from oxidation (Aviram et al. 1998; Fuhrman et al. 2004; Khan-Merchant et al.

Fig. 2 The effects of PON1 on the oxidative capacity of the human carotid plaque lipid fraction. LT marker (40 μM) was incubated overnight with LE, LE treated with PON1 or with inactive PON1. Results expressed as level of LT-OOH. Student's paired *t*-test was performed for the above statistical analyses. Significant differences from LE treatment * $p < 0.01$

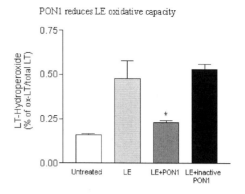

2002; Lusis, 2000), and the enhancement of HDL-mediated cholesterol efflux from macrophages (Rosenblat et al. 2005).

In our experimental system, PON1 protecting effects were most efficient in preventing fatty acid oxidation, rather than nucleic acid oxidation, as indicated by oxidation of LT and LTG synthetic markers. Such protection was not obtained when LE was treated with inactive PON1 (Fig. 2). Since PON1 is known to diminish levels of oxidized lipids, one may speculate that the reduced potential of the PON-treated LE to oxidize LDL, macrophages and exogenous markers is due to PON1's capability to reduce the hydroperoxide content of the human plaque, as we indeed demonstrated by the decrease shown in the peroxide level in PON1-treated LE (Tavori et al. 2008). Thus, the reduction in LE's potency to induce oxidation, due to decreased plaque peroxide concentrations, may account, at least in part, for PON1's anti-atherogenic effects. PON1 also affects other plaque constituents, such as oxysterol composition. Surprisingly, we observed a significant, 3-fold increment in 7-ketocholesterol (7-keto-ch) levels upon PON1 treatment of LE. Other oxysterols such as 7α-hydroxycholesterol (7α-OH-ch), 7β-hydroxycholesterol (7β-OH-ch) and 27-hydroxycholesterol (27OH-ch) remained unchanged. 7-Keto-ch is an abundant oxysterol in human atherosclerotic carotid plaques, and it is widely regarded as pro-atherogenic. It was shown to induce apoptosis of human vascular smooth muscle cells (SMC), to inhibit migration of human SMC and induce inflammation, but under certain conditions an anti-oxidative effect of 7-keto-ch was recently described (Vejux et al. 2008).

7-Keto-ch can potentially inhibit the rate-limiting enzyme in cholesterol biosynthesis, 3-hydroxy-3-methyl-glutaryl-CoA (HMG-CoA) reductase (Lyons and Brown, 1999), and to serve as an inhibitor for 11β-hydroxysteroid dehydrogenase type 1 (11β-HSD1) (Wamil et al. 2008). The immediate question arising thus is whether the robust increase in 7-keto-ch could have originated directly from a reaction of PON1 with 7α- and/or 7β-cholesterol hydroperoxide (7-OOH-ch). 7-Keto-ch may be formed from 7-OOH-ch through a number of pathways, including: (1) a disproportionate reaction in which two molecules of hydroperoxide convert one molecule to 7-OH-ch, while the other one is oxidized to 7-keto-ch and yields a singlet oxygen ("Russell mechanism") (Sheng, 1991), (2) transformation of 7-OOH-ch to a 7-alkoxyl radical catalyzed by bivalent ferrous ion (the alkoxyl radical can react with an additional 7-OOH-ch to form 7-keto-ch and 7-OH-ch [chain termination reaction]) (Murphy et al. 2008), and finally (3) 7-OOH-ch can be dissociated via a hydride ion rearrangement to form 7-keto-ch and a water molecule (March, 1985). In order to test the above possibilities, we synthesized and purified 7-OOH-ch. The pure 7-OOH-ch was subsequently incubated with rePON1, and the resulting reaction mixture was analyzed by TLC and LC/MS, revealing no significant change in the 7-OOH-ch identity and concentration. Further mechanistic analyses are thus needed to explain the simultaneous effects detected in PON1-treated LE, i.e., a reduction in hydroperoxide level with an augmented 7-ketocholesterol concentration.

3 Paraoxonase 1 Lactonase Anti-Atherogenic Activity

PON1 catalyzes the hydrolysis of multiple compounds. It is speculated that the main PON1 activity is the hydrolysis of lactones (Draganov et al. 2005; Gaidukov et al. 2006; Khersonsky and Tawfik, 2005), as well as the reverse lactonization reaction (Santanam and Parthasarathy, 2007; Teiber et al. 2003), while the arylesterase and organophosphatase activities seem to be ancillary (Ahmed et al. 2002). PON1 is believed to primarily act as a lactonase, based on the evolutionary relationship among the close family members (PON1, PON2 and PON3). PON2 has a limited spectrum of substrates, with a specificity only for lactones, and is also the oldest member of the PON family (Draganov and La Du, 2004; La Du et al. 1999). Relating the structure of these various possible substrates (phosphotriesters, esters and lactones) to their rate of hydrolysis showed that lactone hydrolysis is not dependent on the pKa of the departing group, and unlike all other substrates, lactones seem to differ in their K_m rather than their K_{cat} values (Khersonsky and Tawfik, 2005). HDL-mediated cholesterol efflux reduction correlates with the inhibition of PON1 lactonase activity (Tavori et al. 2008), and normalized PON1 lactonase activity correlates with the levels of HDL-associated PON1 and reflects the efficiency of catalytic stimulation of HDL (Gaidukov and Tawfik, 2007). This normalized PON1 lactonase activity was recently suggested to serve as an independent risk factor for coronary artery disease. These findings, and the relatively high rates of hydrolysis measured with several lactones as substrates (Khersonsky and Tawfik, 2006), suggests that PON1 is in fact a lactonase. Furthermore, the lactonase hypothesis of PON1 is also based on the ability of plasma purified PON1 (and PON3), not only to hydrolyze lactones, but also to catalyze the reverse reaction of lactonization (Draganov et al. 2005).

3.1 Characterization of Paraoxonase 1 Active Site, in Relation to Its Lactonase Activity

The pathway by which PON1 mediates its anti-atherogenic function and its natural ligand is not known, although it is believed that these ligands are lipids with lactone-like structure. We carried out modeling and docking techniques as useful tools in

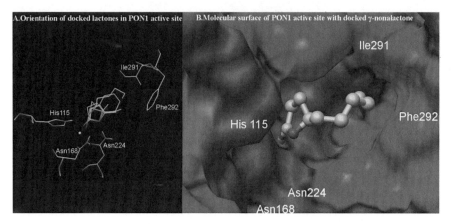

Fig. 3 The orientation of lactones within the PON1 possible binding site. (**a**) Four different lactones (angelica lactone, γ-nonalactone, 2-coumaronone and β-hydroxybutyrolactone) docked into the enzyme active site. (**b**) Molecular surface of PON1 hydrophilic (light) and hydrophobic (dark) zones docked with the γ-nonalactone is shown. The groove to which γ-nonalactone is bound is the predicted enzyme binding site. The carboxylate is directed towards the hydrophilic inner part of the enzyme groove, while the ligand's side chain is facing the hydrophobic part of the groove (*blue*)

structure-based drug design, to predict the fitness of ligands within the protein active site. In such modeling and docking techniques, the evaluated ligands are allowed to interact with the protein 3D structure, enabling the determination of the target and target–ligand complex interactions (Bonete et al. 1996; Goodsell et al. 1996; Khatib et al. 2007).

Employment of such techniques enabled the prediction of the PON1 active site (Tavori et al. 2008) using the enzyme crystallographic structure (Harel et al. 2004), and known lactone ligands (Billecke et al. 2000; Draganov et al. 2005). It was found that all lactones which are known to be catalytically hydrolyzed by PON1 enter the same PON1 groove, and are oriented very similarly one to the other as illustrated in Fig. 3a, taking angelica lactone, γ-nonalactone, 2-coumaronone and β-hydroxybutyrolactone in the enzyme groove as examples. Figure 3 demonstrates that the lactone ring forms a rigid conformation, while the residues attached to the lactone ring have more degrees of freedom, facing the outer (hydrophobic) part on the edge of the enzyme groove (Fig. 3b).This localization of the lactones within the predicted active site is characterized by a specific enzyme–ligand interaction, of Asn168 NH$_2$ residue and the His115 amine forming hydrogen bonds with the ligand carbonyl oxygen, and the Asn224 NH$_2$ residue forming a hydrogen bond with the ligand carboxylate oxygen, thus facilitating the nucleophilic attack on the carbonyl carbon of the lactone, resulting in a ring opening and the generation of carboxylic acid. Therefore, the involvement of His115 residue in the hydrolysis mechanism may be mediated through its direct interaction with lactone substrates in the active site. The catalytic calcium of the enzyme is located at a distance of approximately

1.9 Å from the lactone carbonyl oxygen, which allows for an electrostatic interaction between them, while the nitrogen atoms of Asn270 and Glu53 are both at a distance, over 4 Å, from the lactone oxygen, and as a result they are not likely to create Van der Waals forces or to be directly involved in the catalysis of the lactone hydrolysis (Santanam and Parthasarathy, 2007). These enzyme–ligand interactions resulted in a correlation between the lactones rates of hydrolysis (y) and their calculated docking energy (x), $R^2 = 0.7008$ ($p < 0.0001$):

$$y = -5.7x - 32.189. \tag{1}$$

There was correlation between the rate that PON1 catalyzes lactone hydrolysis relative to the length of the side chain residue (x) on the lactone ring (the lipophilic side chain attached to the γ or δ carbon) versus the calculated docking energy (y) with $R^2 = 0.8498$ ($p < 0.0001$):

$$y = -0.4513x - 6.053, \tag{2}$$

suggesting that the increased hydrophobicity of the side chain improves enzyme–lactone affinity. Another important inverse correlation was found between the rate of hydrolysis of the ligand by PON1 (x) and the dipole moment of the ligand (y), with $R^2 = 0.706$ ($p < 0.0001$):

$$y = -0.211x + 3.643 \tag{3}$$

The dipole moment along the y-axis seems to play an important role and to contribute to the partial charge of the lactone carbonyl carbon, and thus affects the rate of the nucleophilic attack on the carbonyl carbon. Interestingly, the cholesterol lowering drugs, statins, cannot enter the enzyme predicted active site, but rather are positioned on the enzyme's surface, forming nonspecific interactions, or positioned on the edge of the active site, with the lactone ring facing the outside of the groove, far from the catalytic calcium ion. Indeed statins are lactones that are known not to be hydrolyzed by PON1, but rather by PON3 (Draganov et al. 2005). These calculations and docking characteristics enabled us to create a model (Tavori et al. 2008) that may be useful as a tool to better envisage a potential endogenous/exogenous PON1 ligand. Possibly, it could be utilized as an early filter for screening databases of compounds, to predict the ability of specific lactones to be a PON1 substrate.

4 Perspectives and Future Directions

Oxidation of LDL and macrophages in the arterial wall triggers foam cell formation, which leads to atherosclerotic plaque development. The atherosclerotic plaque contains oxidized lipids that are able to induce further oxidation, and thus accelerate atherogenic processes. The identification of the exact endogenous lipid compounds responsible for these oxidative effects, as well as the exact mechanism by which

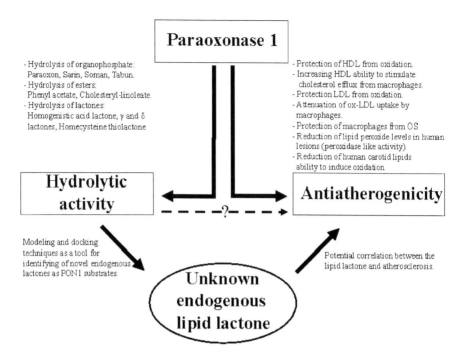

Fig. 4 Suggested mechanisms by which PON1 protects against oxidative stress and accelerated atherogenesis: interactions among PON1's anti-atherogenic functions and its hydrolytic activities

PON1 reduces the lesion lipid fraction atherogenic capacity in association with PON1 lactonase activity, still needs more clarification. Identifying specific endogenous lactone-like oxidized lipids (using also modeling and docking techniques) may help us in a better understanding of how PON1 mediates its anti-atherogenic properties in the lesion, as well as identifying the natural PON1 ligand (Fig. 4).

References

Ahmed, Z.; Ravandi, A.; Maguire, G. F.; Emili, A.; Draganov, D.; La Du, B. N.; Kuksis, A.; Connelly, P. W. Multiple substrates for paraoxonase-1 during oxidation of phosphatidylcholine by peroxynitrite. *Biochem Biophys Res Commun* **290**:391–396; 2002.

Ahmed, Z.; Ravandi, A.; Maguire, G. F.; Emili, A.; Draganov, D.; La Du, B. N.; Kuksis, A.; Connelly, P. W. Apolipoprotein A-I promotes the formation of phosphatidylcholine core aldehydes that are hydrolyzed by paraoxonase (PON-1) during high density lipoprotein oxidation with a peroxynitrite donor. *J Biol Chem* **276**:24473–24481; 2001.

Aviram, M. Review of human studies on oxidative damage and antioxidant protection related to cardiovascular diseases. *Free Radic Res* **33** (Suppl):S85–S97; 2000.

Aviram, M.; Hardak, E.; Vaya, J.; Mahmood, S.; Milo, S.; Hoffman, A.; Billicke, S.; Draganov, D.; Rosenblat, M. Human serum paraoxonases (PON1) Q and R selectively decrease lipid peroxides in human coronary and carotid atherosclerotic lesions: PON1 esterase and peroxidase-like activities. *Circulation* **101**:2510–2517; 2000.

Aviram, M.; Kaplan, M.; Rosenblat, M.; Fuhrman, B. Dietary antioxidants and paraoxonases against LDL oxidation and atherosclerosis development. *Handb Exp Pharmacol* **170**: 263–300; 2005.

Aviram, M.; Maor, I.; Keidar, S.; Hayek, T.; Oiknine, J.; Bar-El, Y.; Adler, Z.; Kertzman, V.; Milo, S. Lesioned low density lipoprotein in atherosclerotic apolipoprotein E-deficient transgenic mice and in humans is oxidized and aggregated. *Biochem Biophys Res Commun* **216**: 501–513; 1995.

Aviram, M.; Rosenblat, M.; Bisgaier, C. L.; Newton, R. S.; Primo-Parmo, S. L.; La Du, B. N. Paraoxonase inhibits high-density lipoprotein oxidation and preserves its functions. A possible peroxidative role for paraoxonase. *J Clin Invest* **101**:1581–1590; 1998.

Billecke, S.; Draganov, D.; Counsell, R.; Stetson, P.; Watson, C.; Hsu, C.; La Du, B. N. Human serum paraoxonase (PON1) isozymes Q and R hydrolyze lactones and cyclic carbonate esters. *Drug Metab Dispos* **28**:1335–1342; 2000.

Bonete, M. J.; Perez-Pomares, F.; Ferrer, J.; Camacho, M. L. NAD-glutamate dehydrogenase from Halobacterium halobium: inhibition and activation by TCA intermediates and amino acids. *Biochim Biophys Acta* **1289**:14–24; 1996.

Costa, L. G.; Vitalone, A.; Cole, T. B.; Furlong, C. E. Modulation of paraoxonase (PON1) activity. *Biochem Pharmacol* **69**:541–550; 2005.

Deakin, S. P.; James, R. W. Genetic and environmental factors modulating serum concentrations and activities of the antioxidant enzyme paraoxonase-1. *Clin Sci (Lond)* **107**:435–447; 2004.

Draganov, D. I.; La Du, B. N. Pharmacogenetics of paraoxonases: a brief review. *Naunyn Schmiedebergs Arch Pharmacol* **369**:78–88; 2004.

Draganov, D. I.; Teiber, J. F.; Speelman, A.; Osawa, Y.; Sunahara, R.; La Du, B. N. Human paraoxonases (PON1, PON2, and PON3) are lactonases with overlapping and distinct substrate specificities. *J Lipid Res* **46**:1239–1247; 2005.

Efrat, M.; Aviram, M. Macrophage paraoxonase 1 (PON1) binding sites. *Biochem Biophys Res Commun* **376**:105–110; 2008.

Efrat, M.; Rosenblat, M.; Mahmood, S.; Vaya, J.; Aviram, M. Di-oleoyl phosphatidylcholine (PC-18:1) stimulates paraoxonase 1 (PON1) enzymatic and biological activities: In vitro and in vivo studies. *Atherosclerosis* **202**:461–469; 2008.

Fuhrman, B.; Judith, O.; Keidar, S.; Ben-Yaish, L.; Kaplan, M.; Aviram, M. Increased uptake of LDL by oxidized macrophages is the result of an initial enhanced LDL receptor activity and of a further progressive oxidation of LDL. *Free Radic Biol Med* **23**:34–46; 1997.

Fuhrman, B.; Oiknine, J.; Aviram, M. Iron induces lipid peroxidation in cultured macrophages, increases their ability to oxidatively modify LDL, and affects their secretory properties. *Atherosclerosis* **111**:65–78; 1994.

Fuhrman, B.; Shiner, M.; Volkova, N.; Aviram, M. Cell-induced copper ion-mediated low density lipoprotein oxidation increases during in vivo monocyte-to-macrophage differentiation. *Free Radic Biol Med* **37**:259–271; 2004.

Gaidukov, L.; Rosenblat, M.; Aviram, M.; Tawfik, D. S. The 192R/Q polymorphs of serum paraoxonase PON1 differ in HDL binding, lipolactonase stimulation, and cholesterol efflux. *J Lipid Res* **47**:2492–2502; 2006.

Gaidukov, L.; Tawfik, D. S. The development of human sera tests for HDL-bound serum PON1 and its lipolactonase activity. *J Lipid Res* **48**:1637–1646; 2007.

[20] Glass, C. K.; Witztum, J. L. Atherosclerosis. the road ahead. *Cell* **104**:503–516; 2001.

Goodsell, D. S.; Morris, G. M.; Olson, A. J. Automated docking of flexible ligands: applications of AutoDock. *J Mol Recognit* **9**:1–5; 1996.

Guardiola, F.; Codony, R.; Addis, P. B.; Rafecas, M.; Boatella, J. Biological effects of oxysterols: current status. *Food Chem Toxicol* **34**:193–211; 1996.

Harel, M.; Aharoni, A.; Gaidukov, L.; Brumshtein, B.; Khersonsky, O.; Meged, R.; Dvir, H.; Ravelli, R. B.; McCarthy, A.; Toker, L.; Silman, I.; Sussman, J. L.; Tawfik, D. S. Structure and evolution of the serum paraoxonase family of detoxifying and anti-atherosclerotic enzymes. *Nat Struct Mol Biol* **11**:412–419; 2004.

Ibanez, B.; Vilahur, G.; Badimon, J. J. Plaque progression and regression in atherothrombosis. *J Thromb Haemost* 5(Suppl 1)**:**292–299; 2007.

Jakubowski, H.; Zhang, L.; Bardeguez, A.; Aviv, A. Homocysteine thiolactone and protein homocysteinylation in human endothelial cells: implications for atherosclerosis. *Circ Res* **87:** 45–51; 2000.

Khan-Merchant, N.; Penumetcha, M.; Meilhac, O.; Parthasarathy, S. Oxidized fatty acids promote atherosclerosis only in the presence of dietary cholesterol in low-density lipoprotein receptor knockout mice. *J Nutr* **132:**3256–3262; 2002.

Khatib, S.; Musa, R.; Vaya, J. An exogenous marker: a novel approach for the characterization of oxidative stress. *Bioorg Med Chem* **15:**3661–3666; 2007.

Khatib, S.; Nerya, O.; Musa, R.; Tamir, S.; Peter, T.; Vaya, J. Enhanced substituted resorcinol hydrophobicity augments tyrosinase inhibition potency. *J Med Chem* **50:**2676–2681; 2007.

Khersonsky, O.; Tawfik, D. S. Structure-reactivity studies of serum paraoxonase PON1 suggest that its native activity is lactonase. *Biochemistry* **44:**6371–6382; 2005.

Khersonsky, O.; Tawfik, D. S. The histidine 115-histidine 134 dyad mediates the lactonase activity of mammalian serum paraoxonases. *J Biol Chem* **281:**7649–7656; 2006.

Kriska, T.; Marathe, G. K.; Schmidt, J. C.; McIntyre, T. M.; Girotti, A. W. Phospholipase action of platelet-activating factor acetylhydrolase, but not paraoxonase-1, on long fatty acyl chain phospholipid hydroperoxides. *J Biol Chem* **282:** 100–108; 2007.

La Du, B. N. *Genetic Factors Influencing the Metabolism of Foreign Compounds.* New York: (international encyclopedia of pharmacology and therapeutics), Pergamon Press; 1992.

La Du, B. N.; Aviram, M.; Billecke, S.; Navab, M.; Primo-Parmo, S.; Sorenson, R. C.; Standiford, T. J. On the physiological role(s) of the paraoxonases. *Chem Biol Interact* **119–120:** 379–388; 1999.

Lusis, A. J. Atherosclerosis. *Nature* **407:**233–241; 2000.

Lyons, M. A.; Brown, A. J. 7-Ketocholesterol. *Int J Biochem Cell Biol* **31:**369–375; 1999.

Mackness, B.; Hunt, R.; Durrington, P. N.; Mackness, M. I. Increased immunolocalization of paraoxonase, clusterin, and apolipoprotein A-I in the human artery wall with the progression of atherosclerosis. *Arterioscler Thromb Vasc Biol* **17:**1233–1238; 1997.

Mackness, B.; Quarck, R.; Verreth, W.; Mackness, M.; Holvoet, P. Human paraoxonase-1 overexpression inhibits atherosclerosis in a mouse model of metabolic syndrome. *Arterioscler Thromb Vasc Biol* **26:**1545–1550; 2006.

Marathe, G. K.; Zimmerman, G. A.; McIntyre, T. M. Platelet-activating factor acetylhydrolase, and not paraoxonase-1, is the oxidized phospholipid hydrolase of high density lipoprotein particles. *J Biol Chem* **278:**3937–3947; 2003.

March, J. Advanced Organic Chemistry. New York: Wiley-Interscience; 1985.

Murphy, R. C.; Johnson, K. M. Cholesterol, reactive oxygen species, and the formation of biologically active mediators. *J Biol Chem* **283:**15521–15525; 2008.

Navab, M.; Berliner, J. A.; Watson, A. D.; Hama, S. Y.; Territo, M. C.; Lusis, A. J.; Shih, D. M.; Van Lenten, B. J.; Frank, J. S.; Demer, L. L.; Edwards, P. A.; Fogelman, A. M. The Yin and Yang of oxidation in the development of the fatty streak. A review based on the 1994 George Lyman Duff Memorial Lecture. *Arterioscler Thromb Vasc Biol* **16:**831–842; 1996.

Paravicini, T. M.; Touyz, R. M. NADPH oxidases, reactive oxygen species, and hypertension: clinical implications and therapeutic possibilities. *Diabetes Care* 31 (Suppl 2): S170-S180; 2008.

Parthasarathy, S.; Litvinov, D.; Selvarajan, K.; Garelnabi, M. Lipid peroxidation and decomposition—conflicting roles in plaque vulnerability and stability. *Biochim Biophys Acta* **1781:**221–231; 2008.

Rosenblat, M.; Gaidukov, L.; Khersonsky, O.; Vaya, J.; Oren, R.; Tawfik, D. S.; Aviram, M. The catalytic histidine dyad of high density lipoprotein-associated serum paraoxonase-1 (PON1) is essential for PON1-mediated inhibition of low density lipoprotein oxidation and stimulation of macrophage cholesterol efflux. *J Biol Chem* **281:**7657–7665; 2006.

Rosenblat, M.; Vaya, J.; Shih, D.; Aviram, M. Paraoxonase 1 (PON1) enhances HDL-mediated macrophage cholesterol efflux via the ABCA1 transporter in association with increased HDL binding to the cells: a possible role for lysophosphatidylcholine. *Atherosclerosis* **179**: 69–77; 2005.

Rozenberg, O.; Rosenblat, M.; Coleman, R.; Shih, D. M.; Aviram, M. Paraoxonase (PON1) deficiency is associated with increased macrophage oxidative stress: studies in PON1-knockout mice. *Free Radic Biol Med* **34**:774–784; 2003.

Rozenberg, O.; Shih, D. M.; Aviram, M. Human serum paraoxonase 1 decreases macrophage cholesterol biosynthesis: possible role for its phospholipase-A2-like activity and lysophosphatidylcholine formation. *Arterioscler Thromb Vasc Biol* **23**:461–467; 2003.

Santanam, N.; Parthasarathy, S. Aspirin is a substrate for paraoxonase-like activity: implications in atherosclerosis. *Atherosclerosis* **191**:272–275; 2007.

Sheng, G. D. M. a. X. C. Yields of excited carbonyl species from alkoxyl and from alkylperoxyl radical dismutations. *J Am Chem Soc* **113**:8976–8977; 1991.

Shih, D. M.; Welch, C.; Lusis, A. J. New insights into atherosclerosis from studies with mouse models. *Mol Med Today* **1**:364–372; 1995.

Skoczynska, A. The role of lipids in atherogenesis. *Postepy Hig Med Dosw (Online)* **59**: 346–357; 2005.

Stadler, N.; Stanley, N.; Heeneman, S.; Vacata, V.; Daemen, M. J.; Bannon, P. G.; Waltenberger, J.; Davies, M. J. Accumulation of zinc in human atherosclerotic lesions correlates with calcium levels but does not protect against protein oxidation. *Arterioscler Thromb Vasc Biol* **28**:1024–1030; 2008.

Stocker, R.; Keaney, J. F., Jr. Role of oxidative modifications in atherosclerosis. *Physiol Rev* **84**:1381–1478; 2004.

Szuchman, A.; Aviram, M.; Musa, R.; Khatib, S.; Vaya, J. Characterization of oxidative stress in blood from diabetic vs. hypercholesterolaemic patients, using a novel synthesized marker. *Biomarkers* **13**:119–131; 2008.

Szuchman, A.; Aviram, M.; Soliman, K.; Tamir, S.; Vaya, J. Exogenous N-linoleoyl tyrosine marker as a tool for the characterization of cellular oxidative stress in macrophages. *Free Radic Res* **40**:41–52; 2006.

Tavori, H.; Aviram, M.; Khatib, S.; Musa, R.; Nitecki, S.; Hofman, A.; Vaya, J. Human carotid atherosclerotic plaque increases oxidative stress of macrophages and LDL, whereas paraoxonase 1 (PON1) decreases such atherogenic effects. *Free Radic Biol Med* **46**:607–15;2009.

Tavori, H.; Khatib, S.; Aviram, M.; Vaya, J. Characterization of the PON1 active site using modeling simulation, in relation to PON1 lactonase activity. *Bioorg Med Chem* **16**: 7504–7509; 2008.

Teiber, J. F.; Draganov, D. I.; La Du, B. N. Lactonase and lactonizing activities of human serum paraoxonase (PON1) and rabbit serum PON3. *Biochem Pharmacol* **66**:887–896; 2003.

Vaya, J.; Aviram, M.; Mahmood, S.; Hayek, T.; Grenadir, E.; Hoffman, A.; Milo, S. Selective distribution of oxysterols in atherosclerotic lesions and human plasma lipoproteins. *Free Radic Res* **34**:485–497; 2001.

Vejux, A.; Malvitte, L.; Lizard, G. Side effects of oxysterols: cytotoxicity, oxidation, inflammation, and phospholipidosis. *Braz J Med Biol Res* **41**:545–556; 2008.

Wamil, M.; Andrew, R.; Chapman, K. E.; Street, J.; Morton, N. M.; Seckl, J. R. 7-Oxysterols modulate glucocorticoid activity in adipocytes through competition for 11{beta}-hydroxysteroid dehydrogense type 11. *Endocrinology* 149(12):5907–5908; 2008.

Watson, A. D.; Berliner, J. A.; Hama, S. Y.; La Du, B. N.; Faull, K. F.; Fogelman, A. M.; Navab, M. Protective effect of high density lipoprotein associated paraoxonase. Inhibition of the biological activity of minimally oxidized low density lipoprotein. *J Clin Invest* **96**:2882–2891; 1995.

Williams, K. J.; Feig, J. E.; Fisher, E. A. Rapid regression of atherosclerosis: insights from the clinical and experimental literature. *Nat Clin Pract Cardiovasc Med* **5**:91–102; 2008.

The Role of Paraoxonase 1 in the Detoxification of Homocysteine Thiolactone

Hieronim Jakubowski

Abstract The thioester homocysteine (Hcy)-thiolactone, product of an error-editing reaction in protein biosynthesis, forms when Hcy is mistakenly selected by methionyl-tRNA synthetase. Accumulating evidence suggests that Hcy-thiolactone plays an important role in atherothrombosis. The thioester chemistry of Hcy-thiolactone underlies its ability to form isopeptide bonds with protein lysine residues, which impairs or alters protein function and has pathophysiological consequences including activation of an autoimmune response and enhanced thrombosis. Mammalian organisms, including human, have evolved the ability to eliminate Hcy-thiolactone. One such mechanism involves paraoxonase 1 (PON1), which has the ability to hydrolyze Hcy-thiolactone. This article outlines Hcy-thiolactone pathobiology and reviews evidence documenting the role of PON1 in minimizing Hcy-thiolactone and *N*-Hcy-protein accumulation.

Keywords Autoantibodies · Atherosclerosis · CBS · Fibrinogen · Hyperhomocysteinemia · Homocysteine thiolactone · Immune activation · MTHFR · Paraoxonase/thiolactonase · Protein *N*-homocysteinylation · Thrombosis

1 Homocysteine Metabolism

In mammals, including humans, homocysteine (Hcy) is formed from methionine (Met) as a result of cellular methylation reactions (Mudd et al., 2001). In this pathway Met is first activated by ATP to yield *S*-adenosylmethionine (AdoMet). As a result of the transfer of its methyl group to an acceptor, AdoMet is converted to *S*-adenosylhomocysteine (AdoHcy). Enzymatic hydrolysis of AdoHcy is the only known source of Hcy in the human body. Levels of Hcy are regulated by remethylation to Met, catalyzed by Met synthase (MS), and transsulfuration to cysteine, the first step of which is catalyzed by cystathionine β-synthase (CBS). The remethylation requires vitamin B_{12} and 5,10-methyl-tetrahydrofolate (CH_3-THF), generated

H. Jakubowski (✉)
Department of Biochemistry and Biotechnology, University of Life Sciences, Poznań, Poland
e-mail: jakubows@umdnj.edu

S.T. Reddy (ed.), *Paraoxonases in Inflammation, Infection, and Toxicology*, Advances in Experimental Medicine and Biology 660, DOI 10.1007/978-1-60761-350-3_11,
© Humana Press, a part of Springer Science+Business Media, LLC 2010

Fig. 1 The metabolism and pathophysiology of Hcy-thiolactone. *N*-Hcy-Fbg, *N*-Hcy-LDL – *N*-homocysteinylated forms of fibrinogen and low density lipoprotein, respectively (adapted from (Chwatko et al., 2007)). See text for discussion

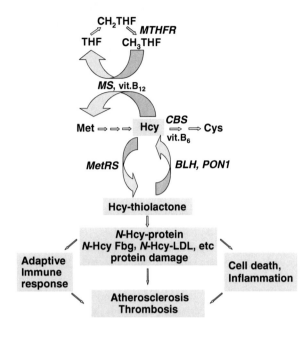

by 5,10-methylene-THF reductase (MTHFR). The transsulfuration requires vitamin B_6 (Fig. 1).

Hcy is also metabolized to the thioester Hcy-thiolactone in an error-editing reaction in protein biosynthesis when Hcy is mistakenly selected in place of Met by methionyl-tRNA synthetase (MetRS) (Fig. 2) (Jakubowski, 2005c, 2000a, 2001a,

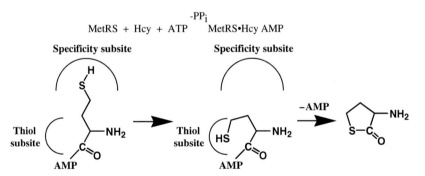

Fig. 2 The formation of Hcy-thiolactone catalyzed by MetRS. During protein biosynthesis Hcy is often mistakenly selected in place of Met by methionyl-tRNA synthetase (MetRS) and activated with ATP to form Hcy-AMP (*upper panel*). The misactivated Hcy is not transferred to tRNA but converted to Hcy-thiolactone in an error-editing reaction (*lower panel*) (adapted from (Jakubowski, 2001))

b, c, 2004, 2005a, Jakubowski and Goldman, 1993). The flow through the Hcy-thiolactone pathway increases when the remethylation or transsulfuration reaction is impaired by genetic alterations of enzymes, such as CBS (Jakubowski, 1991, 1997; 2002a; 2007), MS (Jakubowski, 1991, 2002a), and MTHFR (Jakubowski, 2006), or by inadequate supply of co-factors, e.g. CH_3-THF (Jakubowski, 1997, 2000; Jakubowski et al., 2000). As will be discussed in a greater detail elsewhere in this chapter, Hcy-thioactone is hydrolyzed to Hcy by paraoxonase 1 (PON1) (Fig. 1).

2 Pathophysiology of Hyperhomocysteinemia

Among pathophysiological manifestations of genetic hyperhomocysteinemia, which include mental retardation, ectopia lentis, and osteoporosis, vascular complications, including increased thrombosis, remain the major cause of morbidity and mortality in untreated patients (Kluijtmans et al., 1999; Mudd et al., 1985; Rosenblatt and Fenton, 2001; Yap et al., 2001). McCully observed advanced arterial lesions in children with inborn errors in Hcy metabolism and proposed a hypothesis that Hcy causes vascular disease (McCully, 1969). Even mild hyperhomocysteinemia, quite prevalent in the general population, is associated with an increased risk of vascular events (Clarke et al., 2007; Shishehbor et al., 2008).

Patients suffering from hyperhomocysteinemia improve upon vitamin B therapy, which lowers plasma Hcy levels. For example, Hcy lowering by vitamin B supplementation improves vascular outcomes in CBS-deficient patients, which suggests that Hcy plays a causal role in atherothrombosis. Specifically, untreated CBS-deficient patients suffer one vascular event per 25 patient-years (Mudd et al., 1985) while treated CBS-deficient patients suffer only one vascular event per 263 patient-years (relative risk 0.091, $p < 0.001$) (Yap et al., 2001). Hcy-lowering therapy started early in life also prevents brain disease from severe MTHFR deficiency (Rosenblatt and Fenton, 2001; Strauss et al., 2007).

Furthermore, lowering plasma Hcy by vitamin B supplementation also improves cognitive function in the general population (Durga et al., 2007). High-risk stroke (Lonn et al., 2006; Spence et al., 2005) but not myocardial infarction patients (Bonaa et al., 2006; Lonn et al., 2006) benefit from lowering of plasma Hcy by vitamin B supplementation. These findings suggest that Hcy plays a greater role in stroke than in myocardial infarction, a suggestion consistent with the observations that in untreated CBS-deficient patients cerebrovascular incidents are eight times more frequent than myocardial infarctions (Mudd et al., 1985). These findings also suggest that Hcy plays a more pronounced role in brain compared with heart pathophysiology. Studies of genetic and nutritional hyperhomocysteinemia in animal models provide additional support for a causal role of Hcy in atherothrombosis (Lentz, 2005).

3 The Hcy-Thiolactone Hypothesis

A preponderance of biochemical and genetic data suggest that elevated Hcy promotes a proatherothrombotic phenotype. Proposed mechanisms underlying Hcy pathobiology include protein modification by Hcy-thiolactone, oxidative stress, inflammation and autoimmune response, endothelial dysfunction, and thrombosis (Jakubowski, 2004, 2007; Lentz, 2005).

The Hcy-thiolactone hypothesis (Jakubowski, 1997) states that a pathway initiated by metabolic conversion of Hcy to Hcy-thiolactone catalyzed by methionyl-tRNS synthetase (Fig. 2) (Jakubowski and Goldman, 1993) contributes to Hcy pathobiology (Fig. 1) (Jakubowski, 2004, 2006, 2007). Hcy-thiolactone is a reactive metabolite that causes protein N-homocysteinylation through the formation of amide bonds with protein lysine residues (Fig. 3), which impairs or alters the protein's function (Jakubowski, 1997, 1999). Plasma Hcy-thiolactone and N-linked protein Hcy(N-Hcy-protein), originally discovered in vitro in human fibroblasts and endothelial cells (Jakubowski, 1997, 1999, 2000; Jakubowski et al., 2000), occur in the human body (Chwatko and Jakubowski, 2005a, b; Jakubowski, 2000; Jakubowski et al., 2000), and are greatly elevated under conditions predisposing to atherothrombosis, such as hyperhomocysteinemia caused by mutations in *CBS* or *MTHFR* gene in humans or a high-Met diet in mice (Chwatko et al., 2007; Glowacki and Jakubowski, 2004; Jakubowski, 2001, 2002b; Jakubowski et al., 2008). N-Hcy-protein accumulates in atherosclerotic lesions in ApoE-deficient mice, and the accumulation increases in animals fed a high methionine diet (Perla-Kajan et al., 2008).

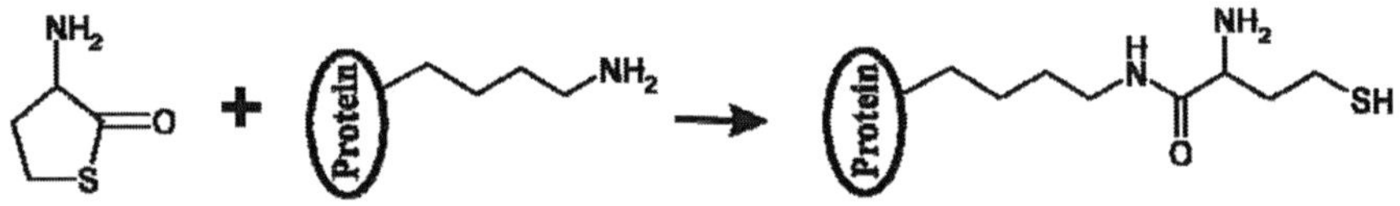

Fig. 3 N-Hcy-protein forms in a reaction of Hcy-thiolactone with a protein lysine residue

4 Toxicity of Hcy-Thiolactone

Accumulating evidence from human, animal, and tissue culture studies shows that Hcy-thiolactone is involved in pathophysiology (reviewed in (Jakubowski, 2004, 2005, 2006, 2007)). Chronic treatments of animals with Hcy-thiolactone cause pathophysiological changes similar to those observed in human genetic hyperhomocysteinemia. For instance, Hcy-thiolactone infusions or Hcy-thiolactone-supplemented diet produce atherosclerosis in baboons (Harker et al., 1974) or rats (Endo et al., 2006) while treatments with Hcy-thiolactone cause developmental abnormalities in chick embryos, including optic lens dislocation (Maestro de las Casas et al., 2003), characteristic of CBS-deficient human patients (Mudd et al.,

2001; 1985). A recent study shows that plasma Hcy-thiolactone levels are associated with the development and progression of diabetic macrovasculopathy (Gu et al., 2008).

Hcy-thiolactone induces apoptotic death in cultured human vascular endothelial (Kerkeni et al., 2006; Mercie et al., 2000) and promyeloid cells (Huang et al., 2001) and placental trophoblasts (Kamudhamas et al., 2004), and inhibits insulin signaling in rat hepatoma cells (Najib and Sanchez-Margalet, 2005). Hcy-thiolactone also induces endoplasmic reticulum (ER) stress and unfolded protein response (UPR) in retinal epithelial cells (Roybal et al., 2004). Furthermore, Hcy-thiolactone is more toxic to cultured cells than Hcy itself (Huang et al., 2001; Kamudhamas et al., 2004; Kerkeni et al., 2006; Mercie et al., 2000; Roybal et al., 2004).

5 Consequences of Protein *N*-Homocysteinylation by Hcy-Thiolactone

Cellular physiology can be impacted by Hcy-thiolactone-mediated protein modification, which changes the primary protein sequence, disrupts protein folding, and creates altered proteins with newly acquired interactions. Small changes in amino acid sequence caused by Hcy incorporation have the potential to create misfolded protein aggregates. Indeed, *N*-Hcy-proteins do have a propensity to form protein aggregates (Jakubowski, 1999). The appearance of misfolded/aggregated proteins in the ER activates a signaling pathway, the UPR, that, when overwhelmed, leads to cell death via apoptosis (Lawrence de Koning et al., 2003; Lentz, 2005). These pathways are induced by treatments of cultured cells and mice with excess Hcy (Hossain et al., 2003), which is metabolized to Hcy-thiolactone (Jakubowski, 1997, 2007; Jakubowski et al., 2000). Hcy-thiolactone is more effective than Hcy in inducing ER and UPR (Roybal et al., 2004). In this scenario the formation of N-Hcy-proteins leads to the UPR and induction of the apoptotic pathway. In humans, Hcy incorporation into proteins triggers an autoimmune response (Jakubowski, 2005b) and increases vascular inflammation (Bogdanski et al., 2007)(Fig. 2), known modulators of atherogenesis (Libby, 2006).

Hcy incorporation is detrimental to protein function. For example, lysine oxidase (Liu et al., 1997), trypsin (Jakubowski, 1999), MetRS (Jakubowski, 1999), and PON1 (Ferretti et al., 2003) are inactivated by *N*-homocysteinylation with Hcy-thiolactone. *N*-Homocysteinylation of albumin (Glowacki and Jakubowski, 2004) and cytochrome *c* (Perla-Kajan et al., 2007) impairs their redox function. *N*-Hcy-proteins (Jakubowski, 1999, 2000), including N-Hcy-LDL (Naruszewicz et al., 1994), tend to form aggregates in vitro. *N*-Hcy-LDL, but not native LDL, induces cell death in human endothelial cells (Ferretti et al., 2004), a finding consistent with the inherent toxicity of protein aggregates (Stefani, 2004).

Plasma *N*-linked Hcy protein is correlated with plasma total Hcy in humans (Jakubowski, 2000, 2002b; Jakubowski et al., 2008). Hcy-thiolactone and *N*-Hcy-protein levels are greatly elevated by CBS or MTHFR deficiency in humans

(Chwatko et al., 2007; Jakubowski et al., 2008). In cultured human cells, CBS deficiency orantifolate drugs such as aminopterin (Jakubowski, 1997) increase the accumulation of Hcy-thiolactone and N-Hcy-protein, whereas supplementation with folic acid decreases the levels of these metabolites (Jakubowski et al., 2000). As will be discussed in greater detail in the final section of this chapter, the accumulation of N-Hcy-protein is also inhibited by the Hcy-thiolactonase activity of PON1 (Fig. 1) (Jakubowski, 2000; Jakubowski et al., 2000).

6 Prothrombotic Properties of N-Hcy-Fibrinogen

Fibrinogen undergoes facile N-homocysteinylation by Hcy-thiolactone in vitro (Jakubowski, 1999, 2000) and in vivo in humans (Jakubowski, 2002b; Jakubowski et al., 2008). Sauls et al. (Sauls et al., 2006) showed that clots formed from Hcy-thiolactone-treated normal human plasma or fibrinogen lyse slower than clots from untreated controls. Some of the lysine residues susceptible to N-homocysteinylation are close to tissue plasminogen activator and plasminogen binding, or plasmin cleavage, sites, which can explain abnormal characteristics of clots formed from N-Hcy-fibrinogen (Sauls et al., 2006). The detrimental effects of elevated plasma tHcy on clot permeability and resistance to lysis in humans are consistent with a mechanism involving fibrinogen modification by Hcy-thiolactone (Undas et al., 2006). Furthermore, CBS-deficient patients have significantly elevated plasma levels of prothrombotic N-Hcy-fibrinogen (Jakubowski et al., 2008), which explains increased atherothrombosis observed in these patients (Mudd et al., 1985).

7 Autoimmunogenicity of N-Hcy-Protein

Atherosclerosis is now widly recognized as a chronic inflammatory disease that involves both innate and adaptive immunity (Libby, 2006). Like other modified proteins, N-Hcy-proteins elicit an autoimmune response in humans, manifested by the induction of IgG autoantibodies directed against Nϵ-Hcy-Lys epitopes. This response is enhanced in stroke and coronary artery disease (CAD) patients, suggesting that it is a general feature of atherosclerosis (Undas et al., 2005, 2004, 2006). Elevated levels of anti-N-Hcy-protein IgG autoantibodies are a consequence of elevated levels of N-Hcy-protein observed in CAD patients (Yang et al., 2006).

The involvement of an autoimmune response against N-Hcy-protein in CAD is supported by the findings that lowering plasma Hcy by folic acid supplementation lowers anti-N-Hcy-protein autoantibodies levels in control subjects but not in patients with CAD (Undas et al., 2006). These findings suggest that, once accumulated, the antigens causing the antibody response, i.e. N-Hcy proteins, persist and that chronic protein damage caused by N-homocysteinylation cannot be easily reversed in CAD patients. Furthermore, these findings also suggest that while primary Hcy-lowering intervention by vitamin supplementation is beneficial, secondary intervention may be ineffective, and may explain at least in part the failure of vitamin therapy to lower cardiovascular events in myocardial infarction patients (Bonaa et al., 2006; Lonn et al., 2006).

8 The Role of PON1 in Elimination of Hcy-Thiolactone

Because Hcy-thiolactone is linked to human pathophysiology (Fig. 1), it is not surprising that the human body evolved the ability to eliminate Hcy-thiolactone. Indeed, we found that an high-density lipoprotein (HDL)-associated enzyme, Hcy-thiolactonase/paraoxonase 1 (PON1) is able to hydrolyze the toxic metabolite Hcy-thiolactone in human serum (Domagała et al. 2006; Jakubowski, 2000; Jakubowski et al., 2001; Lacinski et al., 2004). More recently, Hcy-thiolactonase/bleomycin hydrolase (BLH) was found to hydrolyze Hcy-thiolactone intracellularly (Zimny et al., 2006). Another mechanism of Hcy-thiolactone elimination involves clearance by the kidney (Chwatko and Jakubowski, 2005).

PON1 is synthesized exclusively in the liver and carried on HDL in the circulation. Owing to its ability to detoxify organophosphate insecticides and nerve gases, PON1 has been studied in the field of toxicology since the 1960 s. More recent studies have implicated PON1 in the pathogenesis of cardiovascular disease. For example, PON1-deficient mice are more susceptible to a high-fat diet-induced atherosclerosis than wild-type littermates (however, the animals do not develop atherosclerosis on a normal chow diet) (Shih et al., 1998). *PON1* transgenic mice (carrying three copies of the human *PON1*) are less susceptible to atherosclerosis (Tward et al., 2002). In vitro studies indicate that HDL from PON1-deficient animals does not prevent LDL oxidation, whereas HDL from *PON1* transgenic animals protects LDL against oxidation more effectively than HDL from wild-type mice. However, the biochemical basis for the putative antioxidative function of PON1 is unclear and its pathophysiologically relevant lipid-related substrate is not known (Vos, 2008).

Hcy-thiolactone, ubiquitously present in living organisms, including humans, is a natural substrate of PON1, which therefore should be more appropriately called Hcy-thiolactonase (HTase) (Jakubowski, 2000). The conclusion that PON1 is an HTase is based on proteomic and genetic evidence. For example, purified human HTase has a molecular weight and *N*-terminal amino acid sequence identical to those of human PON1 protein. Sera from PON1-deficient mice are also deficient in HTase activity (Table 1) (Jakubowski, 2000, 2001). HTase activity is absent in serum from chicken (Jakubowski, 2000, 2001), in which the *PON1* gene is known to be absent. HTase activity of PON1 requires calcium for activity and stability, and is inhibited by isoleucine and the antiarthritic drug D-penicillamine (Jakubowski, 2000). Human *PON1* has genetic polymorphisms, e.g. *PON1-M55L, PON1-R192Q* (Jarvik et al., 2000), which are responsible for about 10-fold inter-individual variation in HTase activity (Domagała et al. 2006; Jakubowski et al., 2001; Lacinski et al., 2004). High

Table 1 Sera from *PON1–/–* mice are deficient in Hcy-thiolactonase (HTase) activity (Jakubowski, 2000, 2001; Shih et al., 1998)

Genotype	HTase (units)	Paraoxonase (units)	Arylesterase (units)
PON1+/+ (n = 6)	13.9 ± 0.1	87.9 ± 2.6	119.0 ± 2.4
PON1–/– (n = 6)	2.2 ± 0.4	0.3 ± 0.03	15.9 ± 1.8

HTase activity is associated with L55 and R192 alleles more frequent in blacks than in whites, whereas low HTase activity is associated with M55 and Q192 alleles, more frequent in whites than in blacks (Jakubowski et al., 2001).

In vitro studies show that HTase activity of PON1 prevents Hcy-thiolactone and *N*-Hcy-protein accumulation in cultures of human endothelial cells (Jakubowski et al., 2001, 2000) (Table 2). The high HTase activity form of PON1 affords better protection against protein *N*-homocysteinylation than the low activity form in human serum (Jakubowski et al., 2001) (Fig. 4). High Hcy-thiolactonase activity in rabbit serum protects serum proteins against *N*-homocysteinylation (Fig. 4) (Jakubowski et al., 2001) and could account for the observation that infusions with Hcy-thiolactone failed to produce atherosclerosis in these animals (Donahue et al., 1974). In humans, HTase activity of PON1 is negatively correlated with plasma

Table 2 HDL inhibits Hcy-thiolactone and *N*-Hcy-protein accumulation in cultures of human vascular endothelial cells (Jakubowski et al., 2000)

Culture conditions	Hcy-thiolactone (pmol)	*N*-Hcy-protein (pmol)
10 μM Hcy	120 ± 15	5.5 ± 0.3
10 μM Hcy, 1 mg/mL HDL	25 ± 3	1.9 ± 0.2

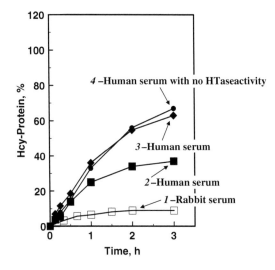

Fig. 4 High HTase activity protects against *N*-Hcy-protein accumulation in human and animal sera. In humans, *L55* and *R192* are high HTase activity, whereas *M55* and *Q192* are low HTase activity *PON1* alleles (adapted from (Jakubowski et al., 2001))

total Hcy (Lacinski et al., 2004) and predicts cardiovascular disease (Domagała et al. 2006). In mice, Hcy is a negative regulator of PON1 expression (Robert et al., 2003).

Recent development of highly sensitive HPLC-based assays facilitated examination of Hcy-thiolactone (Chwatko et al., 2007; Chwatko and Jakubowski, 2005, 2005) and *N*-Hcy-protein (Jakubowski, 2008) metabolism in vivo both in humans and in mice. These assays also allowed examination of the role of PON1 in Hcy-thiolactone and *N*-Hcy-protein metabolism. We found that in C57BL/6 J mice fed a normal diet ($n = 8$), plasma and urinary Hcy-thiolactone levels were 3.7 ± 2.1 nM and 136 ± 22 nM (Table 2), respectively. These levels are similar to plasma and urinary Hcy-thiolactone levels in humans (Chwatko et al., 2007).

In mice fed a high-Met diet ($n = 14$), Hcy-thiolactone concentrations increased 4- to 25-fold, to $13.8.0 \pm 4.8$ and 3490 ± 3780 nM, in plasma and urine, respectively. Plasma and urinary tHcy levels increased 17.3-fold (to 51.8 ± 22.7 µM) and 30-fold (to 1360 ± 840 µM) (Table 3), respectively, compared to mice fed a normal chow diet (Chwatko et al., 2007).

Table 3 Urinary and plasma Hcy-thiolactone and tHcy concentrations in mice (Chwatko et al., 2007) (mean $\pm$ SD)

Diet (6 weeks)	Urinary Hcy-thiolactone (nM)	Urinary tHcy (µM)	Plasma Hcy-thiolactone (nM)	Plasma tHcy (µM)
Control ($n = 8$)	136 ± 22	45 ± 14	3.7 ± 2.1	3.0 ± 1.5
High-Met ($n = 8$)	3490 ± 3780	1360 ± 840	13.8 ± 4.8	51.8 ± 22.7

We then measured levels of plasma Hcy-thiolactone and plasma *N*-Hcy-protein in *PON1–/–* and *PON1+/+* mice fed a normal chow diet or a high-Met diet for up to 18 weeks. The levels of Hcy-thiolactone and *N*-Hcy were then normalized to Hcy levels. We found that the normalized plasma Hcy-thiolactone levels (Hcy-thiolactone/Hcy ratios) were similar in *PON1–/–*and *PON1+/+* mice fed a normal chow diet (Table 4). In animals fed a high-Met diet for 2, 8, and 18 weeks, which increases Hcy-thiolactone and Hcy accumulation (Table 3), the Hcy-thiolactone/Hcy ratios did not differ between *PON1–/–* and *PON1+/+* mice.

Because the bulk of Hcy-thiolactone is excreted in urine (Chwatko et al., 2007; Chwatko and Jakubowski, 2005), we also measured urinary Hcy-thioactone in

Table 4 Plasma Hcy-thiolactone/Hcy ratios in *PON1+/+* and PON1–/– mice fed a chow diet and a high-Met diet for 2, 8 and 18 weeks (H. Jakubowski, D. Shih, unpublished data)

Genotype	Plasma (Hcy-thiolactone/Hcy) $\times$ 1000			
	0	2 weeks	8 weeks	18 weeks
PON1+/+	0.67 ± 0.05	0.66 ± 0.29	0.94 ± 0.32	0.17 ± 0.08
PON1–/–	0.52 ± 0.08	0.59 ± 0.27	0.89 ± 0.17	0.17 ± 0.08

Table 5 Urinary Hcy-thiolactone/Hcy ratios in *PON1+/+* and *PON1–/–* mice fed a chow diet and a high-Met diet for 2, 8 and 18 weeks (H. Jakubowski, D. Shih, unpublished data)

	Urinary (Hcy-thiolactone/Hcy) × 1000		
Genotype	0	2 weeks	18 weeks
PON1+/+	9.3 ± 2.7	45 ± 17	98 ± 74
PON1–/–	29.4 ± 15.0*	87 ± 86	102 ± 133

*$p = 0.02$ for *PON1–/–* vs. *PON1+/+*

PON1–/– and *PON1+/+* mice fed a normal diet or a high-Met diet. We found that urinary Hcy-thiolactone/Hcy ratios were significantly elevated in *PON1–/–* mice relative to *PON1+/+* mice fed a normal chow diet (Table 5). The Hcy-thiolactone/Hcy ratios also increased in *PON1–/–* relative to *PON1+/+* mice fed a high-Met diet, but the difference did not reach statistical significance.

The reaction of Hcy-thiolactone with proteins (Jakubowski, 1997, 1999; Jakubowski et al., 2000) could lead to an apparent lack of Hcy-thiolactone accumulation in mice fed a high-Met diet. To examine this possibility, we measured plasma *N*-Hcy-protein levels in WT and *PON1–/–* mice. Plasma *N*-Hcy-protein levels were normalized to plasma Hcy levels (*N*-Hcy-protein/Hcy ratio). We found that there was no difference in *N*-Hcy-protein/Hcy ratios between *PON1–/–* and *PON1+/+* mice fed a normal chow diet (0.28 ± 0.21 vs. 0.33 ± 0.21). However, the *N*-Hcy-protein/Hcy ratios were significantly higher in *PON1–/–* mice compared with *PON1+/+* mice fed a high-Met diet for 2 weeks (1.21 ± 1.19 vs. 0.25 ± 0.11, $p = 0.027$) and 8 weeks (0.98 ± 0.20 vs. 0.72 ± 0.19, $p = 0.022$) (Table 6). The differences between *PON1–/–* and *PON1+/+* were the highest at 2 weeks, diminished at 8 weeks, and became insignificant at 18 weeks (0.16 ± 0.07 vs. 0.13 ± 0.03) (Table 6). Taken together, these results show that PON1 controls the accumulation of Hcy-thiolactone in mice and that PON1 protects against transient elevation of *N*-Hcy-protein in mice fed a hyperhomocysteinemic high-Met diet.

Table 6 Plasma *N*-Hcy-protein/Hcy ratios in *PON1+/+* and PON1–/– mice fed a chow diet and a high-Met diet for 2, 8 and 18 weeks (H. Jakubowski, D. Shih, unpublished data)

	Plasma *N*-Hcy-protein/Hcy			
Genotype	0	2 weeks	8 weeks	18 weeks
PON1+/+	0.33 ± 0.21	0.25 ± 0.11	0.72 ± 0.19	0.13 ± 0.03
PON1–/–	0.28 ± 0.05	1.21 ± 1.19*	0.98 ± 0.20**	0.16 ± 0.07

* $p = 0.027$, **$p = 0.022$ for PON1–/– vs. *PON1+/+*

Leptin administration in male Wistar rats lowers plasma activities of PON1 (Beltowski et al., 2003), including HTase activity (Beltowski et al., 2008), and increases plasma *N*-Hcy-protein levels, but has no effect on plasma Hcy levels (Beltowski et al., 2008). Co-treatment of rats with a synthetic agonist to liver X receptor, T0901317, prevents leptin-induced decrease in HTase activity of PON1 and increase in *N*-Hcy-protein accumulation in rat serum. Control experiments show that T0901317 has no effect on HTase activity and *N*-Hcy-protein accumulation in animals not receiving leptin. These results suggest that PON1 controls *N*-Hcy-protein accumulation in rats (Beltowski et al., 2008).

9 Conclusion

Despite advances in our understanding of cardiovascular disease, coronary heart disease is still the major cause of mortality in industrial nations. Traditional risk factors such as hypertension, diabetes, hyperlipidemia, and smoking do not accurately predict cardiovascular events. Thus identification of novel risk factors and their mechanisms of action has important public health implications. Hcy is a novel risk factor for the development of cardiovascular disease (Clarke et al., 2007; Shishehbor et al., 2008). Studies of severe genetic hyperhomocysteinemia in humans and genetic and nutritional hyperhomocysteinemia in animal models show that Hcy plays a causal role in atherothrombosis. Metabolic conversion of Hcy to Hcy-thiolactone and inadvertent protein modification by Hcy-thiolactone (which induces pathophysiological responses, such as an autoimmune activation and increased thrombosis) contribute to proatherogenic changes in the cardiovascular system. Chronic activation of these processes in nutritional or genetic deficiencies in Hcy metabolism can cause vascular disease. The ability to hydrolyze Hcy-thiolactone and minimize the accumulation of *N*-Hcy-protein is likely to contribute to the cardioprotective function of PON1.

References

Beltowski J, Wojcicka G, Jamroz A. Leptin decreases plasma paraoxonase 1 (PON1) activity and induces oxidative stress: the possible novel mechanism for proatherogenic effect of chronic hyperleptinemia. Atherosclerosis. 2003 Sep;170(1):21–29.

Beltowski J, Wojcicka G, Jamroz-Wisniewska A, Marciniak A. Liver X receptor agonist, T0901317, normalizes plasma PON1 activity and reduces protein homocysteinylation in rats with experimental hyperleptinemia. 3rd International Conference on Paraoxonases; 2008; Los Angeles, CA. 2008.

Bogdanski P, Pupek-Musialik D, Dytfeld J, Lacinski M, Jablecka A, Jakubowski H. Plasma homocysteine is a determinant of tissue necrosis factor-alpha in hypertensive patients. Biomed Pharmacother. 2007 Dec 3;62:360–365

Bonaa KH, Njolstad I, Ueland PM, Schirmer H, Tverdal A, Steigen T, et al. Homocysteine lowering and cardiovascular events after acute myocardial infarction. N Engl J Med. 2006 Apr 13;354(15):1578–1588.

Chwatko G, Boers GH, Strauss KA, Shih DM, Jakubowski H. Mutations in methylenetetrahydrofolate reductase or cystathionine beta-synthase gene, or a high-methionine diet, increase homocysteine thiolactone levels in humans and mice. Faseb J. 2007 Jun;21(8):1707–1713.

Chwatko G, Jakubowski H. The determination of homocysteine-thiolactone in human plasma. Anal Biochem. 2005 Feb 15;337(2):271–277.

Chwatko G, Jakubowski H. Urinary excretion of homocysteine-thiolactone in humans. Clin Chem. 2005 Feb;51(2):408–415.

Clarke R, Lewington S, Sherliker P, Armitage J. Effects of B-vitamins on plasma homocysteine concentrations and on risk of cardiovascular disease and dementia. Curr Opin Clin Nutr Metab Care. 2007 Jan;10(1):32–39.

Domagała TB, Łacinski M, Trzeciak WH, Mackness B, Mackness MI, Jakubowski H. The correlation of homocysteine-thiolactonase activity of the paraoxonase (PON1) protein with coronary heart disease status. Cell Mol Biol (Noisy-le-grand). 2006;52(5):4–10.

Donahue S, Struman JA, Gaull G. Arteriosclerosis due to homocyst(e)inemia. Failure to reproduce the model in weanling rabbits. Am J Pathol. 1974 Nov;77(2):167–163.

Durga J, van Boxtel MP, Schouten EG, Kok FJ, Jolles J, Katan MB, et al. Effect of 3-year folic acid supplementation on cognitive function in older adults in the FACIT trial: a randomised, double blind, controlled trial. Lancet. 2007 Jan 20;369(9557):208–216.

Endo N, Nishiyama K, Otsuka A, Kanouchi H, Taga M, Oka T. Antioxidant activity of vitamin B6 delays homocysteine-induced atherosclerosis in rats. Br J Nutr. 2006 Jun;95(6): 1088–1093.

Ferretti G, Bacchetti T, Marotti E, Curatola G. Effect of homocysteinylation on human high-density lipoproteins: a correlation with paraoxonase activity. Metabolism. 2003 Feb;52(2): 146–151.

Ferretti G, Bacchetti T, Moroni C, Vignini A, Nanetti L, Curatola G. Effect of homocysteinylation of low density lipoproteins on lipid peroxidation of human endothelial cells. J Cell Biochem. 2004 May 15;92(2):351–360.

Glowacki R, Jakubowski H. Cross-talk between Cys34 and lysine residues in human serum albumin revealed by N-homocysteinylation. J Biol Chem. 2004 Mar 19;279(12):10864–10871.

Gu W, Lu J, Yang G, Dou J, Mu Y, Meng J, et al. Plasma homocysteine thiolactone associated with risk of macrovasculopathy in Chinese patients with type 2 diabetes mellitus. Adv Ther. 2008 Sep;25(9):914–924.

Harker LA, Slichter SJ, Scott CR, Ross R. Homocystinemia. Vascular injury and arterial thrombosis. N Engl J Med. 1974 Sep 12;291(11):537–543.

Hossain GS, van Thienen JV, Werstuck GH, Zhou J, Sood SK, Dickhout JG, et al. TDAG51 is induced by homocysteine, promotes detachment-mediated programmed cell death, and contributes to the development of atherosclerosis in hyperhomocysteinemia. J Biol Chem. 2003 Aug 8;278(32):30317–30327.

Huang RF, Huang SM, Lin BS, Wei JS, Liu TZ. Homocysteine thiolactone induces apoptotic DNA damage mediated by increased intracellular hydrogen peroxide and caspase 3 activation in HL-60 cells. Life Sci. 2001 May 11;68(25):2799–2811.

Jakubowski H. Proofreading in vivo: editing of homocysteine by methionyl-tRNA synthetase in the yeast Saccharomyces cerevisiae. Embo J. 1991 Mar;10(3):593–598.

Jakubowski H. Metabolism of homocysteine thiolactone in human cell cultures. Possible mechanism for pathological consequences of elevated homocysteine levels. J Biol Chem. 1997 Jan 17;272(3):1935–1942.

Jakubowski H. Protein homocysteinylation: possible mechanism underlying pathological consequences of elevated homocysteine levels. Faseb J. 1999 Dec;13(15):2277–2283.

Jakubowski H. Homocysteine thiolactone: metabolic origin and protein homocysteinylation in humans. J Nutr. 2000 Feb;130(2S Suppl):377S–381S.

Jakubowski H. Calcium-dependent human serum homocysteine thiolactone hydrolase. A protective mechanism against protein N-homocysteinylation. J Biol Chem. 2000 Feb 11;275(6): 3957–3962.

Jakubowski H. Biosynthesis and reactions of homocysteine thiolactone. In: Jacobson D, Carmel R (eds.). Homocysteine in Health and Disease. Cambridge, UK: Cambridge University Press; 2001. 21–31.

Jakubowski H. Translational accuracy of aminoacyl-tRNA synthetases: implications for atherosclerosis. J Nutr. 2001 Nov;131(11):2983S–2987S.

Jakubowski H. Protein N-homocysteinylation: implications for atherosclerosis. Biomed Pharmacother. 2001 Oct;55(8):443–447.

Jakubowski H. The determination of homocysteine-thiolactone in biological samples. Anal Biochem. 2002 Sep 1;308(1):112–119.

Jakubowski H. Homocysteine is a protein amino acid in humans. Implications for homocysteine-linked disease. J Biol Chem. 2002 Aug 23;277(34):30425–30428.

Jakubowski H. Molecular basis of homocysteine toxicity in humans. Cell Mol Life Sci. 2004 Feb;61(4):470–487.

Jakubowski H. Accuracy of aminoacyl-tRNA synthetases: Proofreading of amino acids. In: Ibba M, Francklyn C, Cusack S (eds.). The Aminoacyl-tRNA Synthetases. Georgetown, TX: Landes Bioscience/Eurekah.com 2005. 384–396.

Jakubowski H. Anti-N-homocysteinylated protein autoantibodies and cardiovascular disease. Clin Chem Lab Med. 2005;43(10):1011–1014.

Jakubowski H. tRNA Synthetase Editing of Amino Acids. Encyclopedia of Life Sciences. Chichester, UK: John Wiley & Sons, Ltd; 2005. p. http://www.els.net/doi:10.1038/npg.els.0003933.

Jakubowski H. Pathophysiological consequences of homocysteine excess. J Nutr. 2006 Jun;136(6 Suppl):1741S–1749S.

Jakubowski H. The molecular basis of homocysteine thiolactone-mediated vascular disease. Clin Chem Lab Med. 2007;45(12):1704–1716.

Jakubowski H. New method for the determination of protein N-linked homocysteine. Anal Biochem. 2008 Sep 15;380(2):257–261.

Jakubowski H, Ambrosius WT, Pratt JH. Genetic determinants of homocysteine thiolactonase activity in humans: implications for atherosclerosis. FEBS Lett. 2001 Feb 23;491(1–2):35–39.

Jakubowski H, Boers GH, Strauss KA. Mutations in cystathionine beta-synthase or methylenetetrahydrofolate reductase gene increase N-homocysteinylated protein levels in humans. FASEB J. 2008;22:4071–4076.

Jakubowski H, Goldman E. Synthesis of homocysteine thiolactone by methionyl-tRNA synthetase in cultured mammalian cells. FEBS Lett. 1993 Feb 15;317(3):237–240.

Jakubowski H, Zhang L, Bardeguez A, Aviv A. Homocysteine thiolactone and protein homocysteinylation in human endothelial cells: implications for atherosclerosis. Circ Res. 2000 Jul 7;87(1):45–51.

Jarvik GP, Rozek LS, Brophy VH, Hatsukami TS, Richter RJ, Schellenberg GD, et al. Paraoxonase (PON1) phenotype is a better predictor of vascular disease than is PON1(192) or PON1(55) genotype. Arterioscler Thromb Vasc Biol. 2000 Nov;20(11):2441–2447.

Kamudhamas A, Pang L, Smith SD, Sadovsky Y, Nelson DM. Homocysteine thiolactone induces apoptosis in cultured human trophoblasts: a mechanism for homocysteine-mediated placental dysfunction? Am J Obstet Gynecol. 2004 Aug;191(2):563–571.

Kerkeni M, Tnani M, Chuniaud L, Miled A, Maaroufi K, Trivin F. Comparative study on in vitro effects of homocysteine thiolactone and homocysteine on HUVEC cells: evidence for a stronger proapoptotic and proinflammative homocysteine thiolactone. Mol Cell Biochem. 2006 Oct;291(1-2):119–126.

Kluijtmans LA, Boers GH, Kraus JP, van den Heuvel LP, Cruysberg JR, Trijbels FJ, et al. The molecular basis of cystathionine beta-synthase deficiency in Dutch patients with homocystinuria: effect of CBS genotype on biochemical and clinical phenotype and on response to treatment. Am J Hum Genet. 1999 Jul;65(1):59–67.

Lacinski M, Skorupski W, Cieslinski A, Sokolowska J, Trzeciak WH, Jakubowski H. Determinants of homocysteine-thiolactonase activity of the paraoxonase-1 (PON1) protein in humans. Cell Mol Biol (Noisy-le-grand). 2004 Dec;50(8):885–893.

Lawrence de Koning AB, Werstuck GH, Zhou J, Austin RC. Hyperhomocysteinemia and its role in the development of atherosclerosis. Clin Biochem. 2003 Sep;36(6):431–441.

Lentz SR. Mechanisms of homocysteine-induced atherothrombosis. J Thromb Haemost. 2005 Aug;3(8):1646–1654.

Libby P. Inflammation and cardiovascular disease mechanisms. Am J Clin Nutr. 2006 Feb;83(2):456S–4560S.

Liu G, Nellaiappan K, Kagan HM. Irreversible inhibition of lysyl oxidase by homocysteine thiolactone and its selenium and oxygen analogues. Implications for homocystinuria. J Biol Chem. 1997 Dec 19;272(51):32370–32377.

Lonn E, Yusuf S, Arnold MJ, Sheridan P, Pogue J, Micks M, et al. Homocysteine lowering with folic acid and B vitamins in vascular disease. N Engl J Med. 2006 Apr 13;354(15):1567–1577.

Maestro de las Casas C, Epeldegui M, Tudela C, Varela-Moreiras G, Perez-Miguelsanz J. High exogenous homocysteine modifies eye development in early chick embryos. Birth Defects Res A Clin Mol Teratol. 2003 Jan;67(1):35–40.

McCully KS. Vascular pathology of homocysteinemia: implications for the pathogenesis of arteriosclerosis. Am J Pathol. 1969 Jul;56(1):111–128.

Mercie P, Garnier O, Lascoste L, Renard M, Closse C, Durrieu F, et al. Homocysteine-thiolactone induces caspase-independent vascular endothelial cell death with apoptotic features. Apoptosis. 2000 Nov;5(5):403–411.

Mudd SH, Levy HL, Krauss JP. Disorders of transsulfuration. In: Scriver CR, Beaudet AL, Sly WS, Valle D, Childs B, Kinzler KW et al. (eds.). The metabolic and molecular bases of inherited disease. 8th ed. New York: Mc Graw-Hill; 2001. 2007–2056.

Mudd SH, Skovby F, Levy HL, Pettigrew KD, Wilcken B, Pyeritz RE, et al. The natural history of homocystinuria due to cystathionine beta-synthase deficiency. Am J Hum Genet. 1985 Jan;37(1):1–31.

Najib S, Sanchez-Margalet V. Homocysteine thiolactone inhibits insulin-stimulated DNA and protein synthesis: possible role of mitogen-activated protein kinase (MAPK), glycogen synthase kinase-3 (GSK-3) and p70 S6K phosphorylation. J Mol Endocrinol. 2005 Feb;34(1):119–126.

Naruszewicz M, Olszewski AJ, Mirkiewicz E, McCully KS. Thiolation of low density lipoproteins by homocysteine thiolactone causes increased aggregation and altered interaction with cultured macrophages. Nutr Metab Cardiovasc Dis. 1994;4:70–77.

Perla-Kajan J, Marczak L, Kajan L, Skowronek P, Twardowski T, Jakubowski H. Modification by homocysteine thiolactone affects redox status of cytochrome c. Biochemistry. 2007 May 29;46(21):6225–6231.

Perla-Kajan J, Stanger O, Luczak M, Ziolkowska A, Malendowicz LK, Twardowski T, et al. Immunohistochemical detection of N-homocysteinylated proteins in humans and mice. Biomed Pharmacother. 2008 May 2;62(7):473–479.

Robert K, Chasse JF, Santiard-Baron D, Vayssettes C, Chabli A, Aupetit J, et al. Altered gene expression in liver from a murine model of hyperhomocysteinemia. J Biol Chem. 2003 Aug 22;278(34):31504–31511.

Rosenblatt D, Fenton W. Disorders of transsulfuration. In: Scriver C, Beaudet A, Sly W, Valle D, Childs B, Kinzler K, et al. (eds.). The metabolic and molecular bases of inherited disease. 8th ed. New York: Mc Graw-Hill; 2001. 2007–2056.

Roybal CN, Yang S, Sun CW, Hurtado D, Vander Jagt DL, Townes TM, et al. Homocysteine increases the expression of vascular endothelial growth factor by a mechanism involving endoplasmic reticulum stress and transcription factor ATF4. J Biol Chem. 2004 Apr 9;279(15):14844–14852.

Sauls DL, Lockhart E, Warren ME, Lenkowski A, Wilhelm SE, Hoffman M. Modification of fibrinogen by homocysteine thiolactone increases resistance to fibrinolysis: a potential mechanism of the thrombotic tendency in hyperhomocysteinemia. Biochemistry. 2006 Feb 28;45(8):2480–2487.

Shih DM, Gu L, Xia YR, Navab M, Li WF, Hama S, et al. Mice lacking serum paraoxonase are susceptible to organophosphate toxicity and atherosclerosis. Nature. 1998 Jul 16;394(6690):284–287.

Shishehbor MH, Oliveira LP, Lauer MS, Sprecher DL, Wolski K, Cho L, et al. Emerging cardiovascular risk factors that account for a significant portion of attributable mortality risk in chronic kidney disease. Am J Cardiol. 2008 Jun 15;101(12):1741–1746.

Spence JD, Bang H, Chambless LE, Stampfer MJ. Vitamin Intervention For Stroke Prevention trial: an efficacy analysis. Stroke. 2005 Nov;36(11):2404–2409.

Stefani M. Protein misfolding and aggregation: new examples in medicine and biology of the dark side of the protein world. Biochim Biophys Acta. 2004 Dec 24;1739(1):5–25.

Strauss KA, Morton DH, Puffenberger EG, Hendrickson C, Robinson DL, Wagner C, et al. Prevention of brain disease from severe 5,10-methylenetetrahydrofolate reductase deficiency. Mol Genet Metab. 2007 Jun;91(2):165–175.

Tward A, Xia YR, Wang XP, Shi YS, Park C, Castellani LW, et al. Decreased atherosclerotic lesion formation in human serum paraoxonase transgenic mice. Circulation. 2002 Jul 23;106(4): 484–490.

Undas A, Brozek J, Jankowski M, Siudak Z, Szczeklik A, Jakubowski H. Plasma homocysteine affects fibrin clot permeability and resistance to lysis in human subjects. Arterioscler Thromb Vasc Biol. 2006 Jun;26(6):1397–1404.

Undas A, Jankowski M, Twardowska M, Padjas A, Jakubowski H, Szczeklik A. Antibodies to N-homocysteinylated albumin as a marker for early-onset coronary artery disease in men. Thromb Haemost. 2005 Feb;93(2):346–350.

Undas A, Perla J, Lacinski M, Trzeciak W, Kazmierski R, Jakubowski H. Autoantibodies against N-homocysteinylated proteins in humans: implications for atherosclerosis. Stroke. 2004 Jun;35(6):1299–1304.

Undas A, Stepien E, Glowacki R, Tisonczyk J, Tracz W, Jakubowski H. Folic acid administration and antibodies against homocysteinylated proteins in subjects with hyperhomocysteinemia. Thromb Haemost. 2006 Sep;96(3):342–347.

Vos E. Homocysteine levels, paraoxonase 1 (PON1) activity, and cardiovascular risk. JAMA. 2008 Jul 9;300(2):168–169; author reply 9.

Yang X, Gao Y, Zhou J, Zhen Y, Yang Y, Wang J, et al. Plasma homocysteine thiolactone adducts associated with risk of coronary heart disease. Clin Chim Acta. 2006 Feb;364(1–2):230–234.

Yap S, Boers GH, Wilcken B, Wilcken DE, Brenton DP, Lee PJ, et al. Vascular outcome in patients with homocystinuria due to cystathionine beta-synthase deficiency treated chronically: a multicenter observational study. Arterioscler Thromb Vasc Biol. 2001 Dec;21(12):2080–2085.

Zimny J, Sikora M, Guranowski A, Jakubowski H. Protective mechanisms against homocysteine toxicity: the role of bleomycin hydrolase. J Biol Chem. 2006 Aug 11;281(32):22485–22492.

Alteration of PON1 Activity in Adult and Childhood Obesity and Its Relation to Adipokine Levels

Ildikó Seres, László Bajnok, Mariann Harangi, Ferenc Sztanek, Peter Koncsos, and György Paragh

Abstract Obesity as a pathogenic disorder is a predisposing factor for cardiovascular diseases and shows an increasing incidence in the industrialized countries. Adipokines such as leptin, adiponectin and resistin have a great impact on the development of atherosclerosis in obesity. Elevated levels of leptin have been found to be atherogenic whereas decreased levels of adiponectin have been proved to be anti-atherogenic in recent studies. The exact role of resistin in the process of atherosclerosis has so far remained uncertain and controversial. In our recent work, we studied the alteration in human paraoxonase-1 (PON1) activity and adipokine levels; furthermore, we also aimed at identifying the potential correlation between these parameters in this metabolic disorder. We investigated the above-mentioned parameters both in adults and in children, with regard to the emerging role of childhood obesity and to get a clearer view of these factors during a whole lifetime. Investigating the adult population with a broad range of body mass index (BMI) we found significantly increased leptin and significantly decreased adiponectin and resistin levels and PON1 activity in the obese group compared to the lean controls. Adiponectin and resistin levels showed significantly positive correlation, while leptin and BMI showed significantly negative correlation with PON1 activity.

Our findings were similar in childhood obesity: leptin showed significantly negative correlation, while adiponectin showed significantly positive correlation with PON1 activity. We found gender differences in the univariate correlations of leptin and adiponectin levels with PON1 activity in the adult population. In multiple regression analysis, adiponectin proved to be an independent factor of PON1 activity both in childhood and adult obesity, furthermore thiobarbituric acid-reactive substances (TBARS) also proved to be an independent predictor of the enzyme in adults, reflecting the important role of oxidative stress in obesity. Investigating PON 192 Q/R polymorphism by phenotypic distribution (A/B isoenzyme) in obese

I. Seres (✉)

Department of Internal Medicine, Medical and Health Science Centre, University of Debrecen, Debrecen, Hungary

e-mail: seres@internal.med.unideb.hu

S.T. Reddy (ed.), *Paraoxonases in Inflammation, Infection, and Toxicology*, Advances in Experimental Medicine and Biology 660, DOI 10.1007/978-1-60761-350-3_12, © Humana Press, a part of Springer Science+Business Media, LLC 2010

children, we found a significant correlation of PON1 arylesterase activity with leptin and adiponectin levels, and of body fat percentage with PON1 192 B isoenzyme.

According to our studies, these metabolic changes in obesity predispose to the early development of atherosclerosis throughout our whole lifetime. Decreased activity of PON1 and alterations in adipokine levels in childhood obesity could contribute to an early commencement of this process, detected only later in adulthood by increased cardiovascular morbidity and mortality. Changed levels of leptin, adiponectin, resistin and PON1 activity at all ages, just like 192 Q/R polymorphism determined by phenotypic distribution, may be useful markers beside the general risk factors.

Keywords Adipokines · Leptin · Adiponectin · Resistin · Obesity · Paraoxonase-1 · Childhood

1 Introduction

1.1 Obesity and Atherosclerosis: The Role of Adipokines in Atherogenesis and Lipid Peroxidation

Obesity has been identified as a major health concern, and as its level continues to grow steadily the condition threatens to rank among the biggest causes of premature death in both the industrialized and emerging economies. Obesity is an important determinant of atherosclerosis, but the mechanisms behind it are only partially understood. Intra-abdominal or visceral adiposity plays a fundamental role in the stimulation of hyperglycemia, dyslipidemia and hypertension, collectively termed metabolic syndrome, which is one of the most significant risk factors of cardiovascular diseases (Ford, 2005).

Previously, a direct relation was found between general obesity and markers of *oxidative stress* (Mutlu-Turkoglu et al. 2003). In humans the susceptibility of lipids to oxidative modification is an independent risk factor of coronary heart disease (Holvoet et al. 2001). Furthermore, it was also shown that body mass index (BMI) is one of the strongest predictors of circulating oxidized-LDL concentrations (Holvoet, 2006; Ho et al. 2002). In addition, previous studies have demonstrated significant correlations of urinary isoprostanes and ox-LDL/LDL ratio with BMI (Sjogren et al. 2005). Collectively, these findings indicate that weight is an important determinant of oxidative stress. However, the mechanisms by which obesity per se could induce increased oxidative stress are not well defined. According to one hypothesis, oxidative stress is induced by low-grade systemic inflammation. Other factors that have been discussed in relation to increased oxidative stress in obesity are insulin resistance and lipoprotein abnormalities (Weinbrenner and Schroder, 2006).

Previous studies have revealed that humoral factors secreted by adipose tissue contribute to the metabolic syndrome and cardiovascular diseases (Bełtowski et al. 2008). Adipocytes are more than passive fuel storage sites; they release numerous biologically active peptides. Obesity alters adipocytokine secretion profiles, and some obesity-related disorders, including cardiovascular diseases, are associated with dysfunction of adipose tissue (Katagiri et al. 2007).

Leptin, the product of *ob* gene, is a 16 kDa peptide hormone produced mainly by differentiated adipocytes, although leptin expression has been demonstrated in other tissues such as liver, skeletal muscle and placenta. Leptin acts on the central nervous system suppressing food intake and stimulating energy expenditure. Leptin receptors have been found ubiquitously in the body, indicating a general role of leptin. The concentration of leptin is directly proportional to total body fat, and thus obese subjects have usually higher leptin concentrations than non-obese subjects. However, not only the rare human leptin deficiency, but also resistance to the effect of leptin, which is common in obesity, is associated with weight gain. Numerous observations indicate a correlation between serum leptin and cardiovascular diseases (Wannamethee et al. 2007; Patel et al. 2008).

Leptin contributes to the pathogenesis of atherosclerosis by several mechanisms including the enhancement of oxidative stress. Leptin increases the generation of reactive oxygen species by activating c-Jun amino-terminal kinase, AP-1 and nuclear factor kappa B pathways (Bouloumie et al. 1999). Hyperleptinaemia promotes vascular inflammation, stiffness, calcification and proliferation by increasing oxidative stress. ROS enhances the expression of adhesion molecules, which promote the chemotaxis of monocytes in the vessel wall (Curat et al. 2004). By increasing oxidative stress and activating protein kinase C, leptin also increases the production of proatherogenic lipoprotein lipase (LPL) in macrophages. LPL promotes the retention of lipoproteins in the subendothelial space, favors monocyte adhesion and stimulates the transformation of macrophages into foam cells (Dubey and Hersong, 2006). In rats, leptin administration decreased the activity of paraoxonase, and increased plasma and urinary concentrations of isoprostanes reflecting increased oxidative stress (Beltowski et al. 2003; Beltowski, 2006).

Adiponectin is an adipocyte-specific protein with roles in glucose and lipid homeostasis. A negative correlation between obesity and circulating adiponectin has been shown, and decreased adiponectin concentrations are associated with insulin resistance and hyperinsulinemia (Szmitko et al. 2007). Adiponectin also plays anti-atherogenic and anti-inflammatory roles. Accordingly, adiponectin concentrations are decreased in patients with coronary artery disease. The mechanisms underlying the role of adiponectin in lipid peroxidation may involve the regulation of proteins associated with triglyceride metabolism, including CD36, acyl CoA oxidase, 5/-AMP-activated protein kinase and peroxisome proliferator-activated receptor γ (Meier and Gressner, 2004).

Resistin is another adipocyte-specific protein with unknown physiologic roles. Data obtained from animal models suggest that resistin induces insulin resistance. Human studies demonstrated conflicting results in obesity (Rabe et al. 2008). Some of them found a positive correlation between resistin and BMI. Furthermore, resistin may be involved in the regulation of cell proliferation and in the differentiation of fat cells, and in chronic inflammatory reactions associated with obesity (Meier and Gressner, 2004). The role of resistin in the development of obesity-related vascular diseases in humans is still uncertain (Katagiri et al. 2007).

Obesity is associated with derangements in lipid profile, called atherogenic lipid profile, consisting of elevated ApoB and triglyceride levels and low HDL cholesterol. HDL is able to intervene at different stages of the atherosclerotic process (Rosenson, 2006). The antioxidant effect of HDL is mainly provided by the human paraoxonase enzyme 1 (PON1). PON1 has been shown to protect LDL from oxidative modification by hydrolyzing lipid hydroperoxides and cholesterol ester hydroperoxides (esterase activity), with peroxidase activity that reducing peroxides to their respective hydroxides and thiolactonase activity hydrolyzing homocysteine thiolactone, thus protecting proteins from homocysteinylation (Beltowski, 2005).

Serum PON1 activity has been shown to be decreased in several diseases with disturbed lipid metabolism that are associated with accelerated atherogenesis, e.g., in diabetes mellitus (Mackness and Durrington, 1995), hypercholesterolemia (Tomás et al. 2000), coronary heart disease (Mackness et al. 2004), and in hemodialysis patients (Paragh et al. 1998), and with aging (Seres et al. 2004).

1.2 Relationship of Leptin to PON1

There are only a few studies that have examined the effects of leptin on PON1 activity and none has investigated the correlation between adiponectin, resistin and PON1 activity.

1.2.1 Animal Studies

It was demonstrated that leptin administered for 7 days decreased plasma PON1 activity and induced oxidative stress in rats (Beltowski, 2005). In addition, leptin reduced PON1 activity in the aorta, renal cortex and medulla but not in the heart, lung or liver. Interestingly, leptin decreased PON1 activity only in tissues in which it stimulated oxidative stress. These data suggest that leptin-induced decrease in PON1 in tissues results from excessive ROS production, consistent with a well-known inactivation of the enzyme by oxidative processes which may be involved in atherogenesis in hyperleptinemic obese individuals.

1.2.2 Human Studies

An inverse relation of PON1 with obesity and serum leptin levels has been demonstrated. Ferretti et al. demonstrated that the increase in oxidative stress in HDL and LDL of obese subjects (monitored by the levels of lipid hydroperoxides in HDL and LDL) was associated with a decrease of PON1 activity in isolated HDL (Ferretti et al. 2005). Uzun et al have also shown an inverse correlation between serum leptin levels and PON1 activity in morbid obesity after gastric banding (Uzun et al. 2004). These data suggest that hyperleptinemia could lead to reduced PON1 activity in humans. However, in our previous work we have showed that hyperleptinemia occurring in chronic renal failure was not responsible for decreased paraoxonase activity (Varga et al. 2006). Moreover, we found a positive correlation between leptin and PON1 activity in hemodialysis (HD) patients. This relationship might be explained by the elevated PON1 activity in HD patients with BMI > 25 kg/m^2

compared to the group with BMI < 25 kg/m^2. However, the reason for enhanced PON1 activity in the high BMI group of HD patients remains unknown.

The above-mentioned studies were performed on animal models, on isolated HDL or in HD patients with special metabolic conditions. However, the relation of these adipokines – leptin, adiponectin and resistin – to PON1 has not been clarified in obesity.

The goal of our study was to examine the relationship between adipokines and PON1 activity in obese adults and obese children (when the cardiovascular complications are not yet manifested).

In order to magnify the impact of obesity on the investigated parameters, we selected the adult study population with a broad range of BMI ranging from 19 to 53 with a mean of 34.2 ($\pm$ 7.11) kg/m^2. Patients were divided into three groups according to BMI values, and were age- and sex-matched (Bajnok et al. 2007). Obese (BMI = 28–39.9 kg/m^2) and morbidly obese (BMI > 40 kg/m^2) patients had significantly higher blood pressures and plasma glucose levels and had atherogenic lipid profiles compared to lean subjects (BMI = 20–24.9 kg/m^2).

We found a significant negative correlation between PON1 activity and BMI ($r = -0.503$, $p < 0.001$). According to BMI categories, we found that obese and morbidly obese patients had significantly lower PON1 activity compared to lean subjects (Fig. 1a). Serum leptin concentration was significantly higher in both obese groups compared to the lean group (Fig 1b). Examining the correlation

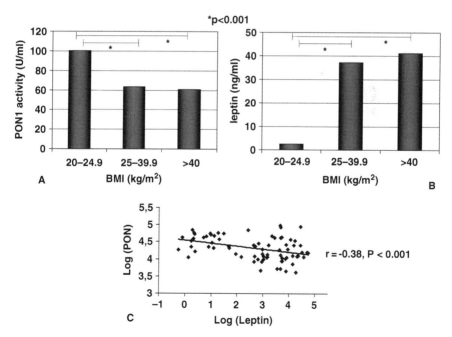

Fig. 1 PON1 paraoxonase activities (**a**) and serum leptin concentrations (**b**) in the different BMI categories, and correlation between PON1 activity and leptin levels (**c**) in adult obesity

between PON1 activity and leptin levels, we found a significant negative correlation between these two parameters, similar to animal studies ($r = -0.38$, $p <$
0.001, Fig 1c). The relationship between leptin and PON1 was also evaluated in
male and female subjects separately, and a stronger correlation was found among
men ($r = -0.50$, $p < 0.01$) than among women ($r = -0.28$, $p = 0.06$) (Bajnok
et al. 2007).

2 Relationship of Adipokines to PON1 in Adult Obesity

2.1 Relationship of Adiponectin to PON1

It was previously demonstrated that adiponectin levels in obese patients were significantly decreased and inversely correlated with both body weight and fat mass
(Díez and Iglesias, 2003). In confirming this well-known fact, we also demonstrated that the obese and morbidly obese patient groups had significantly lower
adiponectin levels compared to the lean group (Fig. 2a). Our results show a negative
and significant correlation between adiponectin levels and BMI ($r = -0.38$, $p <$
0.001) (Fig. 2b). Since several in vivo and in vitro studies have reported that
adiponectin has direct anti-atherogenic effects on the arterial wall (Matsuda et al.

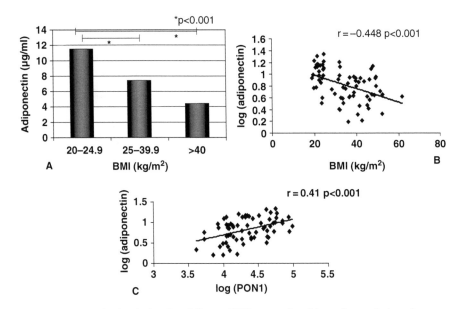

Fig. 2 Adiponectin levels in the different BMI categories (**a**), and correlations between
adiponectin level and BMI (**b**) and between adiponectin level and PON1 activity (**c**) in adult
obesity

2002) and that hypoadiponectinemia was associated with endothelial dysfunction (Shimabukuro et al. 2003), cardiovascular disease (Kumada et al. 2003) and diabetes mellitus (Hotta et al. 2000), we hypothesized that there might be correlations between antioxidant PON1 activities and adiponectin levels. We analyzed the relationship of adiponectin concentration to PON1 activity (Bajnok et al. 2008a). To the best of our knowledge this was the first study to investigate the relationship between PON1 activity and adiponectin levels. We found a strong positive correlation between these two anti-atherogenic factors ($r = 0.41$; $p < 0.001$) (Fig. 2c). There was a stronger correlation between adiponectin and PON1 among males ($r = 0.49$, $p < 0.01$) than among females ($r = 0.37$, $p < 0.05$).

To test if the association between adiponectin and PON1 was independent of anthropometric and other laboratory parameters, we carried out a multiple regression analysis. During this test only adiponectin turned out to be an independent predictor of serum PON1, but none of the other variables that were included in the model (Table 1). The association between adiponectin and PON1 was independent of anthropometric and other parameters, i.e. age, gender, BMI, systolic blood pressure, insulin resistance index by homeostasis model assessment (HOMA-IR), LDL-C, HDL-C and lipid peroxidation (measured by thiobarbituric acid-reactive substances [TBARS]).

Table 1 Multiple regression analysis for PON1 activity as a dependent variable

Variable	β	P
Intercept	4.289	< 0.001
Age	−0.002	0.724
Gender	−0.052	0.629
BMI	0.003	0.746
Systolic BP	0.001	0.928
HOMA-IR	−0.120	0.186
LDL-C	−0.061	0.658
HDL-C	−0.079	0.574
TBARS	−0.032	0.698
Adiponectin	0.252	0.011

BMI: Body mass index; BP: blood pressure; HDL-C: high-density lipoprotein-cholesterol; HOMA-IR: insulin resistance index by homeostasis model assessment; LDL-C: low-density lipoprotein-cholesterol; TBARS: thiobarbituric acid-reactive substances

2.2 Relationship of Resistin to PON1

Resistin levels were higher in the controls than among the obese subjects, with no differences between the obese subgroups (Fig. 3a). The concentrations of resistin were similar in the three groups of PON1 phenotypes, and there were no significant differences in the distribution of PON1 phenotypes among the three BMI groups of

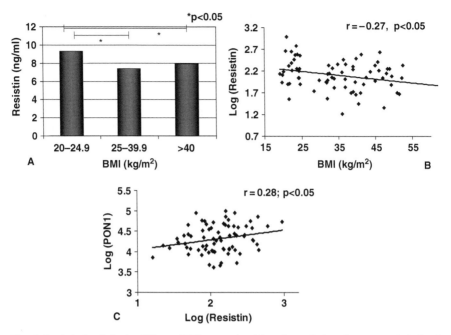

Fig. 3 Resistin levels in the different BMI categories (**a**), and correlations between resistin level and BMI (**b**) and between resistin level and PON1 activity (**c**) in adult obesity

subjects. Impact of BMI categories on the resistin level remained significant even after adjustments for age and gender (Bajnok et al. 2008b).

Univariate correlation analysis showed that in the whole population, serum levels of resistin were correlated negatively with BMI and correlated positively with PON1 activity (Fig. 3b,c).To test if the association of resistin with PON1 existing in the univariate analysis was independent of anthropometric and other laboratory parameters, we carried out multiple regression analyses with PON1 as the dependent variable.

At first, two less well adjusted models (Model *A* and *B*) were constructed in which, beside resistin, the impact of age, sex and BMI were tested (Table 2). In Model *B* HDL-C was also included, since PON1 is associated with a subfraction of HDL. In these models only BMI turned out to be an independent predictor of PON1, explaining the 12.6% of the variance of PON1. We also constructed a more fully adjusted model (Model *C*), applying parameters that are related either to metabolic syndrome (systolic blood pressure, HDL-C, HOMA-IR) and/or lipid peroxidation (LDL-C and TBARS). The reason for including these latter two parameters was that higher levels of cholesterol concentration and lipid peroxidation are associated with enhanced inactivation of PON1 in an interaction between lipid peroxides and the sulfhydryl groups of the enzyme.

In the extended model (Model *C*), BMI ceased to be a significant independent variable of PON1, and of the investigated parameters only TBARS proved to be a

Table 2 Multiple regression analysis for PON1 paraoxonase activity as a dependent variable

Variable	Model A ($R^2 = 0.20$)			Model B ($R^2 = 0.21$)			Model C ($R^2 = 0.22$)		
	B	t	P	β	t	P	β	t	P
Age	0.08	0.67	0.50	0.04	0.34	0.73	0.05	0.37	.071
Gender	−0.09	−0.79	0.44	−0.14	−1.22	0.23	−0.04	−0.27	0.79
BMI	−0.42	−3.87	< 0.001	−0.42	−3.87	< 0.001	0.11	0.39	0.70
Resistin	0.14	1.22	0.23	0.14	1.22	0.23	0.11	0.78	0.44
HDL-C	–	–	0.11	0.82	0.42	0.06	0.34	0.74	
LDL-C	–	–	–	–	–	−0.06	−0.42	0.68	
TBARs	–	–	–	–	–	−0.41	−3.2	0.002	
Systolic BP	–	–	–	–	–	−0.07	−0.35	0.73	
HOMA-IR	–	–	–	–	–	−0.23	−1.54	0.13	

BMI: Body mass index; BP: blood pressure; HDL-C: high-density lipoprotein-cholesterol; HOMA-IR: insulin resistance index by homeostasis model assessment; LDL-C: low-density lipoprotein-cholesterol; TBARS: thiobarbituric acid-reactive substances

predictor of PON1, showing that of the partially related variables, TBARS and BMI, the variance of PON1 was explained better by TBARS than by BMI (Table 2).

To the best of our knowledge, this is the first report of the relationship between resistin and PON1. However, when we tested if the association between resistin and PON1 was independent of anthropometric and other parameters in multiple regression analysis, resistin was not an independent predictor of PON1. In fact, during this multivariate analysis only the negative correlation between PON1 and lipid peroxidation (measured by TBARS) remained significant, and neither the BMI, nor the age, gender, systolic BP, HOMA-IR, LDL-C or HDL-C were significant predictors of PON1 activity.

3 Relationship of Adipokines to PON1 in Childhood Obesity

3.1 Association of Adipokines and PON1 Activity in Childhood Obesity

Genetic polymorphisms in the promoter and coding regions of the *PON1* gene are the main determinants of the enzyme activity, but serum PON1 activity can be modulated by several other factors. Aging and pathologic states such as renal disease, diabetes mellitus, cardiovascular disease, and liver cirrhosis are associated with decreased PON1 activity, and various dietary and lifestyle factors have been reported to influence serum PON1 activity (Seres et al. 2004; Paragh et al. 1998, 1999; Mackness and Durrington, 1995; Ferré et al. 2006). Smoking has been associated with reduced PON1 activity and concentrations in patients with coronary artery disease. On the other hand, lipid-lowering therapy with statins (Tomás et al. 2000;

Paragh et al. 2004), and hormone replacement therapy (Sutherland et al. 2001), have been demonstrated to increase serum PON1 activity.

Since other factors influencing PON1 activity (smoking, concomitant diseases like type 2 diabetes mellitus (DM2), chronic renal failure (CRF), cardiovascular disease (CVD)) can be present in obese adults, we decided to investigate PON1 activity in obese children, where the incidence of these factors are markedly lower. Similar to the results in adult patients, obese children (age: 11.95 $\pm$ 1.61 years, BMI: 28.23 $\pm$ 4.33 kg/m^2) had significantly higher serum leptin levels (43.61 $\pm$ 26.64 vs. 11.69 $\pm$ 14.63 ng/ml; $p < 0.001$) that correlated positively with their body fat percentage ($r = 0.52$; $p < 0.001$). (Body fat percentage (BFP) is a better characterizing parameter for obesity in childhood than BMI.) We found gender differences in leptin levels both in the obese and the age- and gender-matched normal-weight groups, and similarly to other studies girls had significantly higher leptin levels than boys ($p < 0.05$). Adiponectin levels were significantly lower in the obese children group compared to the control group (8.59 $\pm$ 4.39 vs. 12.24 $\pm$ 4.86 μg/ml, $p < 0.001$), as was expected. Obese children had significantly lower PON1 activity (97.31 $\pm$ 21.24 vs. 111.44 $\pm$ 23.52 U/L; $p < 0.01$). Similar to adult obese individuals, we demonstrated an inverse relation between PON1 activity and leptin levels ($r = -0.29$, $p < 0.05$) and a positive relation between PON1 activity and adiponectin concentrations ($r = 0.39$; $p < 0.01$). In order to test whether the associations of PON1 with leptin and adiponectin existing in the univariate analysis were independent of other parameters, we carried out a multiple regression analysis. Adiponectin level also proved to be an independent predictor of PON1 activity after adjusting for age, sex, BFP, leptin and HDL-C in the model (Table 3).

Table 3 Multiple regression analysis for PON1 arylesterase as a dependent variable (model $R^2 = 0.442$, $p < 0.05$)

Independent variable	Regression coefficient	SE of regression coefficient	Standardized coefficient (β)	t	p
Age	1.039	2.513	0.068	0.413	0.68
Sex	13.356	7.168	0.326	1.863	0.08
Leptin	−0.312	0.180	−0.349	−1.734	0.10
Adiponectin	2.151	0.841	0.453	2.557	0.02
HDL-cholesterol	−20.480	15.920	−0.236	−1.286	0.21
Body fat percentage	−0.192	0.526	−0.073	−0.365	0.72

3.2 Correlations Between Adipokine Levels and PON1 Phenotypes

Investigating the phenotypic distribution of PON1 in the obese group, we found differences in the correlations of PON1 activity in the subgroups (Figs. 4 and 5). We could divide the obese children into two groups: 40 children belonging to the

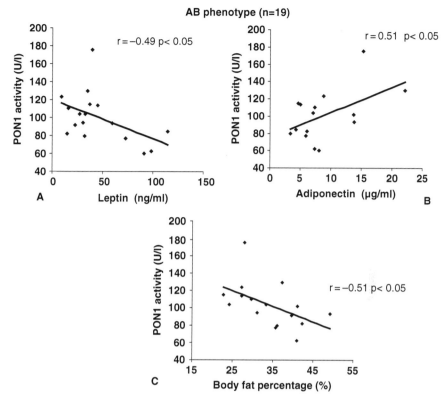

Fig. 4 Correlations between PON1 activity and serum leptin levels (**a**), serum adiponectin levels (**b**) and body fat percentage (BFP) (**c**) in obese children with AB phenotype

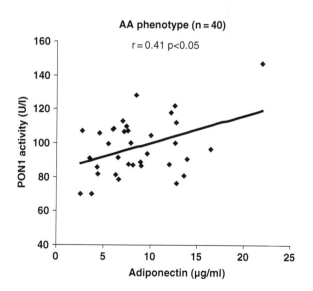

Fig. 5 Correlation between PON1 activity and serum adiponectin level in obese children with AA phenotype

group with AA phenotype and 19 children to the group with AB phenotype; none of the obese children had BB phenotype. We did not find any significant differences in anthropometric and clinical data between the two groups. Obese children with AB phenotype had a significant correlation between PON1 activity (Fig. 4a,b,c) and serum leptin ($r = -0.49$, $p < 0.05$), adiponectin levels ($r = 0.51$, $p < 0.05$) and BFP ($r = -0.51$, $p < 0.05$) whereas children with AA phenotype had a significant correlation between PON1 arylesterase activity and adiponectin levels ($r = 0.41$, $p < 0.05$; Fig. 5) but with no significant correlation observed between PON1 activity and the other investigated parameters.

4 Conclusions

Taken together, in a population with a broad range of BMI that could be divided into three equal groups, we found a significant negative correlation between PON1 activity and BMI ($r = -0.503$ $p < 0.001$) that was independent of age, sex and HDL-C; BMI explained 12.6% of the variance of PON1. Compared to lean subjects, obese patients had significantly higher leptin and lower adiponectin and resistin levels and PON1 activity. Serum PON1 activity showed a negative correlation with leptin ($r = -0.38$, $p < 0.001$), while a positive correlation was shown with adiponectin ($r = 0.41$; $p < 0.001$) and resistin. However, only adiponectin turned out to be an independent predictor of PON1 activity in a multiple regression model in which other factors, such age, gender, BMI, systolic blood pressure, HOMA-IR, LDL-C, HDL-C and a marker of lipid peroxidation were also included, beside the adipokines.

These results were confirmed among children, except resistin that was not investigated.

The low PON1 activity in obese children is a novel and alarming result. Previously we had found that serum PON1 activity significantly decreased with age (Seres et al. 2004). It may mean that the initially higher cardiovascular risk of obese children caused by lower PON1 activity will be even higher with aging. Therefore, screening and treatment of these children is especially important to prevent early manifestations of atherosclerosis.

Our studies suggest that obesity predisposes to accelerated progression of atherosclerosis throughout our whole lifetime. Childhood obesity also demonstrates the importance of this pathologic metabolic state, the consequences of which can be detected only in later adulthood, and that is why the investigation of this population has become so important. Changed levels of leptin, adiponectin, resistin and PON1 activity, just like 192 Q/R polymorphism determined by phenotypic distribution, may be useful markers beside the general risk factors in both adult and childhood obesity.

Acknowledgments This work was supported by a grant from the Hungarian Scientific Research Fund (OTKA K63025), ETT (Medical Research Council) (243/2006) and Hungarian National Office for Research and Technology (OMFB-1613/2006), Hungary.

References

Bajnok L, Seres I, Varga Z et al. (2007) Relationship of endogenous hyperleptinemia to serum paraoxonase 1, cholesteryl ester transfer protein, and lecithin cholesterol acyltransferase in obese individuals. Metabolism 56:1542–1549

Bajnok L, Csongradi E, Seres I et al. (2008a) Relationship of adiponectin to serum paraoxonase-1. Atherosclerosis. 197:363–367

Bajnok L, Seres I, Varga Z et al. (2008b) Relationship of serum resistin level to traits of metabolic syndrome and serum paraoxonase 1 activity in a population with a broad range of body mass index. Exp Clin Endocrinol Diabetes 116:592–599

Beltowski J, Wójcicka G, Jamroz A (2003) Leptin decreases plasma paraoxonase 1 (PON1) activity and induces oxidative stress: the possible novel mechanism for proatherogenic effect of chronic hyperleptinemia. Atherosclerosis 170:21–29

Beltowski J (2005) Protein homocysteinylation: a new mechanism of atherogenesis? Postepy Hig Med Dosw 59:392–404

Beltowski J. (2006) Leptin and atherosclerosis. Atherosclerosis 189:47–60

Beltowski J, Jamroz-Wiśniewska A and Widomska S (2008) Adiponectin and its role in cardiovascular diseases. Cardiovasc Hematol Disord Drug Targets 8:7–46

Bouloumie A, Marumo T, Lafontan M et al. (1999) Leptin induces oxidative stress in human endothelial cells. FASEB J 13:1231–1238

Curat CA, Miranville A, Sengenès C et al. (2004) From blood monocytes to adipose tissue-resident macrophages: induction of diapedesis by human mature adipocytes. Diabetes 53:1285–1292

Díez JJ and Iglesias P (2003) The role of the novel adipocyte-derived hormone adiponectin in human disease. Eur J Endocrinol 148:293–300

Dubey L and Hersong Z (2006) Role of leptin in atherogenesis. Exp Clin Cardiol 11:269–275

Ferré N, Marsillach J, Camps J et al. (2006) Paraoxonase-1 is associated with oxidative stress, fibrosis and FAS expression in chronic liver diseases. J Hepatol 45:51–59

Ferretti G, Bacchetti T, Moroni C et al. (2005) Paraoxonase activity in high-density lipoproteins: a comparison between healthy and obese females. J Clin Endocrinol Metab 90:1728–1733

Ford ES (2005) Risks for all-cause mortality, cardiovascular disease, and diabetes associated with the metabolic syndrome: a summary of the evidence. Diabetes Care 28:1769–1778

Ho RC, Davy K, Davy B et al. (2002) Whole-body insulin sensitivity, low-density lipoprotein (LDL) particle size, and oxidized LDL in overweight, nondiabetic men. Metabolism 51: 1478–1483

Holvoet P, Mertens A, Verhamme P et al. (2001) Circulating oxidized LDL is a useful marker for identifying patients with coronary artery disease. Arterioscler Thromb Vasc Biol 21:844–848

Holvoet P (2006) Obesity, the metabolic syndrome, and oxidized LDL. Am J Clin Nutr 83:1438

Hotta K, Funahashi T, Arita Y et al. (2000) Plasma concentrations of a novel, adipose-specific protein, adiponectin, in type 2 diabetic patients, Arterioscler Thromb Vasc Biol 20:1595–1599

Katagiri H, Yamada T and Oka Y (2007) Adiposity and cardiovascular disorders. Disturbance of the regulatory system consisting of humoral and neuronal signals. Circ Res 101:27–39

Kumada M, Kihara S, Sumitsuji S et al. (2003) Association of hypoadiponectinemia with coronary artery disease in men. Arterioscler Thromb Vasc Biol 23:85–89

Mackness MI, Durrington PN (1995) Paraoxonase: another factor in NIDDM cardiovascular disease. Lancet 346:856

Mackness MI, Durrington PN, Mackness B (2004) The role of paraoxonase 1 activity in cardiovascular disease: potential for therapeutic intervention. Am J Cardiovasc Drugs 4: 211–217

Matsuda M, Shimomura I, Sata M et al. (2002) Role of adiponectin in preventing vascular stenosis: the missing link of adipo-vascular axis. J Biol Chem 277:37487–37491

Meier U and Gressner AM (2004) Endocrine regulation of energy metabolism: review of pathobiochemical and clinical chemical aspects of leptin, ghrelin, adiponectin, and resistin. Clin Chem 50:1511–1525

Mutlu-Turkoglu U, Oztezcan S, Telci A et al. (2003) An increase in lipoprotein oxidation and endogenous lipid peroxides in serum of obese women. Clin Exp Med 2:171–174

Paragh Gy, Seres I, Balogh Z et al. (1998) The serum paraoxonase activity in patients with chronic renal failure and hyperlipidemia. Nephron 80:166–170

Paragh Gy, Asztalos L, Seres I et al. (1999) Serum paraoxonase activity changes in uremic and kidney transplanted patients. Nephron 83:126–131

Paragh G, Töröcsik D, Seres I et al. (2004) Effect of short term treatment with simvastatin and atorvastatin on lipids and paraoxonase activity in patients with hyperlipoproteinaemia. Curr Med Res Opin 20:1321–1327

Patel SB, Reams GP, Spear RM et al. (2008) Leptin: linking obesity, the metabolic syndrome, and cardiovascular disease. Curr Hypertens Rep 10:131–137

Rabe K, Lehrke M, Parhofer KG et al. (2008) Adipokines and insulin resistance. Mol Med 14: 741–751

Rosenson RS (2006) Low high-density lipoprotein cholesterol and cardiovascular disease: risk reduction with statin therapy. Am Heart J 151:556–563

Seres I, Paragh Gy, Deschene E et al. (2004) Study of factors influencing the decreased HDL associated PON1 activity with aging. Exp Gerontol 39:56–66

Shimabukuro M, Higa N, Asahi T et al. (2003) Hypoadiponectinemia is closely linked to endothelial dysfunction in man. J Clin Endocrinol Metab 88:3236–3240

Sjogren P, Basu S, Rosell M et al. (2005) Measures of oxidized low-density lipoprotein and oxidative stress are not related and not elevated in otherwise healthy men with the metabolic syndrome. Arterioscler Thromb Vasc Biol 25:2580–2586

Sutherland W, Manning PJ, de Jong SA et al. (2001) Hormone-replacement therapy increases serum paraoxonase arylesterase activity in diabetic postmenopausal women. Metabolism 50:319–324

Szmitko PE, Teoh H, Stewart DJ et al. (2007) Adiponectin and cardiovascular disease: state of the art Am J Physiol Heart Circ Physiol 292:1655–1663

Tomás M, Sentí M, García-Faria F et al. (2000) Effect of simvastatin therapy on paraoxonase activity and related lipoproteins in familial hypercholesterolemic patients. Arterioscler Thromb Vasc Biol 20:2113–2119

Uzun H, Zengin K, Taskin M et al. (2004) Changes in leptin, plasminogen activator factor and oxidative stress in morbidly obese patients following open and laparoscopic Swedish adjustable gastric banding. Obes Surg 14:659–665

Varga Z, Paragh G, Seres I et al. (2006) Hyperleptinemia is not responsible for decreased paraoxonase activity in hemodialysis patients. Nephron Clin Pract 103:114–120

Wannamethee SG, Tchernova J, Whincup P et al. (2007) Plasma leptin: associations with metabolic, inflammatory and haemostatic risk factors for cardiovascular disease. Atherosclerosis 191:418–426

Weinbrenner T, Schröder H, Escurriol V et al. (2006) Circulating oxidized LDL is associated with increased waist circumference independent of body mass index in men and women. Am J Clin Nutr 83:30–35

Anti-Inflammatory Properties of Paraoxonase-1 in Atherosclerosis

Bharti Mackness, and Mike Mackness

Abstract Atherosclerosis is increasingly recognised as an inflammatory disease. The inflammatory process begins with the oxidation of low-density lipoprotein (LDL) in the artery wall. The ability of high-density lipoprotein (HDL) to inhibit the oxidation of LDL (and cell membranes) and promote macrophage cholesterol efflux through the action of several of its associated proteins, particularly paraoxonase-1 (PON1), reduces the inflammation associated with atherosclerosis. In vivo, in animal models, ablation of the *PON1* gene is pro-inflammatory and pro-atherogenic, while overexpression of human PON1 is anti-inflammatory and anti-atherogenic. In subjects with diabetes mellitus, PON1 is dysfunctional due to glycation, reducing its ability to retard LDL and cell membrane oxidation and contributing to the inflammation typical of diabetes, leading to the excess atherosclerosis common in this disease.

Keywords Paraoxonase-1 · High-density lipoprotein · Inflammation · Oxidation · Lipid hydroperoxides · Paraoxonase-3

1 Inflammation and Atherosclerosis

It is now commonly accepted that atherosclerosis is an inflammatory disease (Ross, 1993). Low-density lipoprotein (LDL) in the sub-intimal space of the artery wall becomes modified by oxidation (ox-LDL). Ox-LDL, in turn, stimulates the production of a variety of cytokines and chemokines by resident macrophages, smooth muscle and endothelial cells in an inflammatory response which results in the upregulation of endothelial cell adhesion molecules, such as vascular cell adhesion molecule-1 and intracellular adhesion molecule-1, which attach circulating monocytes (Lusis, 2000). The attached monocytes are attracted into the intima by the ox-LDL-stimulated production of the chemokine, monocyte chemoattractant protein-1 (MCP-1). The monocytes in the intima differentiate into macrophages that

B. Mackness (✉)
Universitat Rovari i Virgili, Center de Recerca Biomedica
Hospital Universitari de Sant Joan, 43201 Reus, Spain
e-mail: bhartimackness@tiscali.co.uk

S.T. Reddy (ed.), *Paraoxonases in Inflammation, Infection, and Toxicology*, Advances in Experimental Medicine and Biology 660, DOI 10.1007/978-1-60761-350-3_13,
© Humana Press, a part of Springer Science+Business Media, LLC 2010

take up ox-LDL in an unregulated manner to become foam cells, the progenitors of atheroma (Steinberg et al.al. 1989).

2 High-Density Lipoprotein (HDL) and Inflammation

2.1 Proinflammatory HDL

Under certain conditions such as acute infection, anti-inflammatory proteins such as PON1 and clusterin are displaced from HDL by pro-inflammatory proteins such as serum amyloid A and haptoglobin, resulting in anti-inflammatory HDL becoming pro-inflammatory (Ansell et al. 2007). Such HDL has been termed dysfunctional. In certain populations associated with premature or excess atherosclerosis, such as those with diabetes or rheumatoid arthritis, and South Asian immigrants, HDL is largely dysfunctional. In a study comparing dysfunctional HDL in subjects with coronary diseases and controls, Ansell et al. found that dysfunctional HDL was better than HDL cholesterol concentration at distinguishing controls from coronary patients (Ansell et al. 2003).

2.2 Anti-Inflammatory HDL

HDL is anti-inflammatory under normal conditions. HDL can break the inflammatory cycle in several ways:

(i) HDL inhibits the oxidation of LDL (Fig. 1).
(ii) HDL inhibits cytokine-induced endothelial cell adhesion molecule production (Nichols et al. 2005).

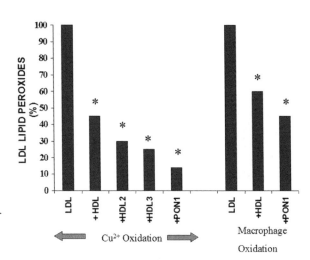

Fig. 1 Inhibition of LDL oxidation by HDL and PON1. Significantly different from LDL, * p < 0.001. Modified from Mackness et al. 2002

(iii) HDL inhibits ox-LDL-induced MCP-1 production by endothelial cells (Navab et al. 1991).

(iv) HDL promotes cholesterol efflux from macrophage foam cells (Aviram and Rosenblat, 2004).

Thus, HDL has the potential to prevent/retard atherosclerosis. This is borne out by the well-known inverse relationship between HDL concentration and coronary disease (Miller and Miller, 1975). In recent years attention has turned to increasing HDL as a way of preventing excess coronary disease. Several pharmacological therapies are currently being investigated, including apolipoprotein mimetics, CETP inhibitors, reconstituted (synthetic) HDL, liver X receptor agonists and PPARα agonists. Although these therapies appear to be efficacious at increasing HDL cholesterol and preventing atherosclerosis development in animal models, they have so far proven less successful in human studies than was hoped (Mackness and Mackness, 2008). This has led to questions as to whether the simplified approach of raising total HDL cholesterol concentration is viable, especially as plasma HDL cholesterol concentration does not reflect its anti-atherosclerotic properties (Mackness and Mackness, 2008).

Many of the anti-inflammatory properties of HDL are centred on its ability to inhibit the oxidation of LDL (Aviram et al, 1998; Mackness et al. 1991; Watson et al. 1995). Several HDL-associated proteins can contribute to this activity (Fig. 2), including paraoxonase-1 and -3, apolipoprotein (apo) A1, lecithin, cholesterol acyltransferase (LCAT) and cholesteryl-ester transfer protein (CETP). Several of these proteins such as PON1, apo A1 and LCAT appear to act synergistically (Fig. 2); however, PON1 activity alone appears to be the major HDL-associated antioxidant. Although in preliminary studies CETP appears to be equally efficient (Fig. 2), these results require further verification. Some authors have claimed that PON1 plays no role in the antioxidant activity of HDL and that this activity is due to the presence of

Fig. 2 Inhibition of LDL oxidation by various components of HDL. Significantly different from LDL, $^*p < 0.01$, $++ p < 0.001$. Modified from Arrol et al. 1996

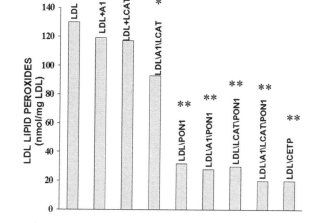

platelet activating factor acetyltransferase (PAFAH) (Marathe et al. 2003). However, these experiments were conducted under conditions likely to inactivate PON1 and we have consistently been unable to find any PAFAH protein associated with HDL (Rodrigo et al. 2001).

3 PON1 and Inflammation

Although hepatic PON1 production can be modulated by inflammatory cytokines (Feingold et al. 1998), a consistent relationship between serum PON1 and cytokine concentration has proven hard to find in humans. In type 2 diabetes mellitus, high high-sensitivity C-reactive protein coupled with low PON1 was predictive of the presence of coronary disease, but there was no significant relationship between the two (Mackness et al. 2006).

In the intima, PON1 can prevent inflammation in two major ways. Firstly, PON1 inhibits the oxidation of LDL, in doing so preventing the ox-LDL-induced upregulation of pro-inflammatory cytokines and chemokines by cells in the vessel wall. Thus, PON1 prevents the ox-LDL-induced upregulation of MCP-1 production by endothelial cells (Watson et al. 1995 and Fig. 3). Interestingly, PON1 cannot prevent pro-inflammatory TNFα induce MCP-1 production by endothelial cells, indicating the specificity of the action of PON1.

Secondly, elegant studies by Aviram and colleagues have indicated that PON1 promotes the mobilisation of cholesterol within macrophages and stimulates the efflux of this cholesterol to HDL (Aviram and Rosenblat, 2004).

Perhaps the most convincing evidence of the anti-inflammatory actions of PON1 comes from in vivo studies in animal models of atherosclerosis. PON1 knockout mice exhibit pro-inflammatory HDL, increased susceptibility of LDL to oxidation

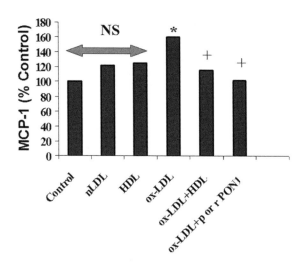

Fig. 3 Inhibition of ox-LDL-induced endothelial cell MCP-1 production by HDL and PON1. N = normal, ox = oxidised, p = purified human, r = recombinant human. Significantly different from control. $^*p < 0.01$. Significantly different from ox-LDL, $+p < 0.001$. Adapted from Mackness et al 2004

and increased atherosclerosis compared to wild-type littermates (Shih et al. 1998). The anti-inflammatory properties of HDL are restored upon addition of purified human PON1. On the other hand, overexpressing human PON1 in mouse models of atherosclerosis reduces systemic inflammation and ox-LDL in both plasma and the artery wall and inhibits the development of atherosclerosis (Mackness et al. 2006a).

In the artery wall of humans, PON1 increases with the development of atherosclerosis, probably as a response to increased oxidative stress (Mackness et al. 1997). In humans and mice, PON1 mRNA expression is restricted to liver, kidney and colon, yet in both these species PON1 protein is found within a large variety of tissues including brain, gut, eye, and many more, indicating a wider anti-inflammatory role for PON1 (Marsillach et al. 2008).

3.1 PON1 and Cell Membrane Oxidation

One recently discovered property of PON1 is its ability to metabolise cell membrane lipid hydroperoxides (Ferretti et al. 2004 and Mastorikou et al. 2008 and Fig. 4). In these studies, erythrocyte ghosts were used as model cell membranes and PON1 was shown to significantly reduce membrane hydroperoxides. PON1 from subjects with either type 1 or type 2 diabetes was significantly less able to metabolise membrane hydroperoxides compared to control PON1. The reason for this difference appears to be the increased glycation of PON1 in the diabetic serum. In subjects with type 2 diabetes, approximately 15% of the PON1 is glycated compared to 7.5% in control subjects (Mastorikou et al. 2008). This increased glycation results in over 50% reduction in PON1 activity towards paraoxon and membrane hydroperoxides (Fig. 5). Interestingly, glyoxidation results in a further reduction in PON1 activity towards both of these substrates (Fig. 5).

The metabolism of cell membrane hydroperoxides may be an additional anti-inflammatory mechanism of PON1, preventing the formation of pro-inflammatory oxidised lipids. Further investigations are required to show whether PON1 can

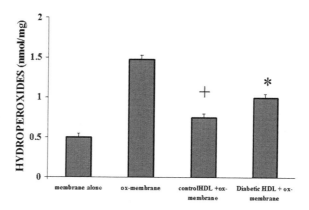

Fig. 4 Metabolism of cell membrane hydroperoxides by PON1. Significantly different from ox-membrane, $+p < 0.01$. Significantly different from control HDL, $*p < 0.05$. Adapted from Mastorikou et al. 2008

Fig. 5 Effect of glycation and glyoxidation on HDL-PON1 activity towards paraoxon and membrane hydroperoxides. Significantly different from HDL, *$p <$ 0.01, **$p < 0.001$. Data redrawn from Mastorikou et al. 2008

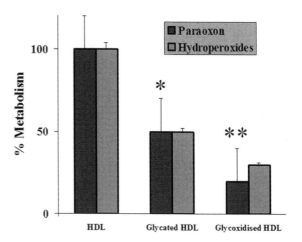

metabolise membrane hydroperoxides of cells of the artery wall. These results also indicate that PON1 from subjects with diabetes is dysfunctional. Previous studies have shown that diabetic PON1 is also less able to metabolise oxidised palmitoyl, arachidonyl phosphatidylcholine, the major oxidised phospholipid found in LDL, than control PON1 (Mastorikou et al. 2006). Taken together, these results indicate that in diabetes, the deranged metabolism of oxidised lipids by PON1 may contribute to the increased inflammation typical of diabetes and the subsequent increased susceptibility to atherosclerosis.

4 Is PON1 Cytotoxic?

In a recent study, we investigated the ability of recombinant human PON1 and PON3 to prevent LDL oxidation, reduce macrophage oxidative stress and promote macrophage cholesterol efflux (Liu et al. 2008). During the experiments investigating cholesterol efflux using the human THP-1 macrophage, we were surprised to find an acute, concentration-dependent cytotoxicity of PON1 (Fig. 6). PON3 was significantly less cytotoxic at the same concentrations. Both proteins were in the same buffer and both associated with DMPC micelles. We could find no other reports of PON1 cytotoxicity and found no cytotoxicity of either recombinant or purified human PON1 to endothelial cells (Mackness et al. 2004). Aviram and colleagues have reported no cytotoxicity of recombinant or purified human PON1 to murine macrophages (Aviram and Rosenblat, 2004). It is therefore possible that in this instance the cytotoxicity is cell-line specific. However, it remains a possibility that under certain conditions PON1 is cytotoxic, as any cytotoxicity not detected would greatly skew studies on the effect of PON1 on macrophage cholesterol efflux and/or oxidative stress. Further studies are warranted to investigate this phenomenon.

Fig. 6 Cytotoxicity of PON1 and PON3 to human macrophages. Significantly different from 0 PON, $*p < 0.01$, $**p < 0.001$. Significantly different from PON1, $+p < 0.001$. Data redrawn from Liu et al. 2008

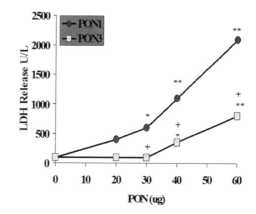

5 Conclusion

There is now strong evidence that the anti-inflammatory properties of HDL are due largely to the actions of PON1 acting alone or in concert with other HDL-associated enzymes in preventing/retarding the oxidation of LDL (and cell membranes), thereby preventing the pro-inflammatory properties of ox-LDL in upregulating endothelial cell chemokines production and in promoting cholesterol efflux from macrophage foam cells. These properties of PON1 result in the inhibition of atherosclerosis development in vivo. Dietary and or pharmaceutical modulation of PON1 to prevent atherosclerosis is now a viable option.

References

Ansell, B.J., Fonarow, G.C., Fogelman, A.M. (2007) The paradox of dysfunctional high-density lipoprotein. Curr Opin Lipidol 18: 427–434.

Ansell, B.J., Navab, M., Hama, S., Kamranpour, N., Fonarow, G., Hough, G. et al. (2003) Inflammatory/antiinflammatory properties of high-density lipoprotein distinguish patients from control subjects better than high-density lipoprotein cholesterol levels and are favourably affected by simvastatin treatment. Circulation 108: 2751–2756.

Arrol, S., Mackness, M.I., Durrington, P.N. (1996) High-density lipoprotein associated enzymes and the prevention of low-density lipoprotein oxidation. Eur J Lab Med 4: 33–38.

Aviram, M., Billecke, S., Sorenson, R., Bisgaier, C., Newton, R., Rosenblat, M., Erogul, J., Hsu, C., Dunlop, C., La Du, B.N (1998) Paraoxonase active site required for protection against LDL oxidation involves its free sulphydryl group and is different from that required for its arylesterase/paraoxonase activities: selective action of human paraoxonase alloenzymes Q and R. Arterioscl Thromb Vasc Biol 10: 1617–1624.

Aviram, M., Rosenblat, M. (2004) Paraoxonases 1, 2, and 3, oxidative stress, and macrophage foam cell formation during atherosclerosis development. Free Radic Biol Med 37: 1304–16.

Feingold, K.R., Memon, R.A., Moser, A.H., Grunfeld, C. (1998) Paraoxonase activity in the serum and hepatic mRNA levels decrease during the acute phase response. Atherosclerosis 139: 307–315.

Ferretti, G., Bacchetti, T., Busni, D., Rabini, R.A., Curatola, G. (2004) Protective effect of paraoxonase activity in high-density lipoproteins against erythrocyte membranes peroxidation: a comparison between healthy subjects and type 1 diabetic patients. J Clin Endocrinol Metab 89: 2957–2962.

Liu, Y., Mackness, B., Mackness, M., (2008) Comparison of the ability of paraoxonases 1 and 3 to attenuate the in vitro oxidation of low-density lipoprotein and reduce macrophage oxidative stress. Free Radic Biol Med 45:743–748

Lusis, A.J. (2000) Atherosclerosis. Nature 407: 233–241.

Mackness, B., Mackness, M. (2009) HDL - Why all the fuss? Ann Clin Biochem- 46:5–7

Mackness, B., Hine, D., Liu, Y. Mastorikou, M., Mackness, M. (2004) Paraoxonase 1 inhibits oxidised LDL-induced MCP-1 production by endothelial cells. BBRC 318: 680–683.

Mackness, B., Hine, D., McElduff, P., Mackness, M. (2006) High C-reactive protein and low paraoxonase 1 in diabetes as risk factors for coronary heart disease. Atherosclerosis 185: 396–401.

Mackness, B., Hunt, R., Durrington, P.N., Mackness, M.I. (1997) Increased immunolocalisation of paraoxonase, clusterin and apolipoprotein AI in the human artery wall with progression of atherosclerosis. Arterioscler Thromb Vasc Biol 17: 1233–1238.

Mackness, B., Quarck, R., Verreth, W., Mackness, M., Holvoet, P. (2006a) Human paraoxonase-1 overexpression inhibits atherosclerosis in a mouse model of metabolic syndrome. Arterioscler Thromb Vasc Biol 26:1545–1550.

Mackness, M.I., Arrol, S. & Durrington, P.N. (1991) Paraoxonase prevents accumulation of lipoperoxides in low-density lipoprotein. FEBS Letts 286: 152–154.

Mackness, MI., Durrington, PN., Mackness, B. (2002) The role of paraoxonase in lipid metabolism. In: Costa, LG., Furlong, CE (eds.). Paraoxonase (PON1) in health and disease. Norwell, MA: Kluwer.

Marathe, G.K., Zimmerman, G.A., McIntyre, T.M. (2003) Platelet-activating factor acetylhydrolase, and not paraoxonase-1, is the oxidized phospholipid hydrolase of high density lipoprotein particles. J Biol Chem 278: 3937–3947.

Marsillach, J., Mackness, B., Mackness, M., Riu, F., Beltran, R., Joven, J., Camps, J. (2008) Immunohistochemical analysis of paraoxonases- 1, 2, and 3 expression in normal mouse tissues. Free Rad Biol Med 45: 146–157.

Mastorikou, M., Mackness, B., Liu, Y., Mackness, M. (2008) Glycation of paraoxonase-1 inhibits its activity and impairs the ability of high-density lipoprotein to metabolise membrane lipid hydroperoxides. Diabetic Med 25: 1049–1055.

Mastorikou, M., Mackness, M., Mackness, B. (2006) Defective metabolism of oxidised-phospholipid by high-density lipoprotein from people with type 2 diabetes. Diabetes 55: 3099–3103.

Miller G.J., Miller N.E. (1975) Plasma high-density lipoprotein concentration and the development of ischaemic heart disease. Lancet 1: 16–19.

Navab, M., Imes, S.S., Hama, S.Y., Hough, G.P., Ross, L.A., Bork, R.W., Valente, A.J., Berliner, J.A., Drinkwater, D.C., Laks, H. & Fogelman, A.M. (1991) Monocyte transmigration induced by modification of low density lipoprotein in cocultures of human aortic wall cells is due to induction of monocyte chemotactic protein 1 synthesis and is abolished by high density lipoprotein. J Clin Invest 88: 2039–2046.

Nicholls, S.J., Dusting, G.J., Cutri, B. Bao, S., Drummond, G.R., Rye K-A., Barter, P.J. (2005) Reconstituted high-density lipoproteins inhibit the acute pro-oxidant and proinflammatory vascular changes induced by periarterial collar in normocholesterolemic rabbits. Circulation 111: 1543–1550.

Rodrigo, L., Mackness, B., Durrington, P.N., Hernandez, A., Mackness, M.I. (2001) Hydrolysis of platelet-activating factor by human serum paraoxonase. Biochem J 354: 1–7.

Ross, R. (1993) The pathogenesis of atherosclerosis: a perspective for the 1990's. Nature 362: 801–809.

Shih, D.M., Gu, L., Xia Y-R., Navab, M., Li, W-F., Hama, S., Castellani, L.W., Furlong, C.E., Costa, L.G., Fogelman, A.M., Lusis, A.J. (1998) Mice lacking serum paraoxonase are susceptible to organophosphate toxicity and atherosclerosis. Nature 394: 284–287.

Steinberg, D., Parthasarathy, S., Carew, T.E., Khoo, J.C. and Witztum, J.L. (1989) Beyond cholesterol modifications of low-density lipoprotein that increase its atherogenicity. New Engl J Med 320: 915–924.

Watson, A.D., Berliner, J.A., Hama, S.Y., La Du, B.N., Fault, K.F., Fogelman, A.M., Navab, M. (1995)Protective effect of high density lipoprotein associated paraoxonase - Inhibition of the biological activity of minimally oxidised low-density lipoprotein. J Clin Invest 96: 2882–2891.

Paraoxonase 1 Interactions with HDL, Antioxidants and Macrophages Regulate Atherogenesis – A Protective Role for HDL Phospholipids

Michal Efrat and Michael Aviram

Abstract Macrophage cholesterol accumulation and foam cell formation is the hallmark of early atherogenesis. In addition to macrophages, at least three more major players regulate atherosclerosis development; paraoxonase 1 (PON1), antioxidants, and HDL. PON1 is an HDL-associated lactonase which posses antioxidant and anti-atherogenic properties. PON1 protects against macrophage-mediated LDL oxidation, and increases HDL binding to macrophages which, in turn, stimulates HDL's ability to promote cholesterol efflux. These two major anti-atherogenic properties of HDL (and of PON1) require, at least in part, macrophage binding sites for HDL-associated PON1. Indeed, PON1, as well as HDL-associated PON1, specifically binds to macrophages, leading to anti-atherogenic effects. Macrophage PON1 binding sites may thus be a target for future cardioprotection therapy. Studying the interactions among PON1, antioxidants, and macrophages can thus assist in achieving appropriate treatment and prevention of atherosclerosis.

Keywords Paraoxonase 1 · HDL · Macrophages · Atherosclerosis · Antioxidants · Phospholipids · Di-oleoyl phosphatidylcholine (PC-18:1) · Olive oil

1 Atherosclerosis

1.1 Atherosclerosis Development

Atherosclerotic lesions contain large numbers of immune cells, particularly macrophages and T cells. Furthermore, the disease is associated with systemic immune responses and inflammation (National Cholesterol Education Program, 1993). During recent years, experiments in gene-targeted mice have provided mechanistic evidence in support of the hypothesis that immune mechanisms are indeed involved in atherosclerosis (National Cholesterol Education Program, 1993).

M. Efrat (✉)
The Lipid Research Laboratory, Technion Faculty of Medicine, the Rappaport Family Institute for Research in the Medical Sciences and Rambam Medical Center, Haifa, Israel
e-mail: michalef@gmail.com

S.T. Reddy (ed.), *Paraoxonases in Inflammation, Infection, and Toxicology*, Advances in Experimental Medicine and Biology 660, DOI 10.1007/978-1-60761-350-3_14,
© Humana Press, a part of Springer Science+Business Media, LLC 2010

Because high plasma concentration of cholesterol, in particular low-density lipoprotein (LDL) cholesterol, is one of the principal risk factors for atherosclerosis, (National Cholesterol Education Program, 1993), the process of atherogenesis has been considered by many to consist largely of the accumulation of lipids within the artery wall; however, it is much more than that. Despite changes in lifestyle and the use of new pharmacologic approaches to lower plasma cholesterol concentrations (Scandinavian Simvastatin Survival Study Group, 1994), cardiovascular disease continues to be the principal cause of death in the US, Europe, and much of Asia (Breslow, 1997; Braunwald EShattuck Lecture, 1997). In fact, the lesions of atherosclerosis represent a series of highly specific cellular and molecular responses that can best be described, in aggregate, as an inflammatory disease (Idem, 1976; Ross, 1986; Ross and Glomset, 1973). The lesions of atherosclerosis occur principally in large and medium-sized elastic and muscular arteries and can lead to ischemia of the heart, brain, or extremities, resulting in infarction. They may be present throughout a person's lifetime. In fact, the earliest type of lesion, the so-called fatty streak, which is common in infants and young children (Napoli et al. 1997), is a pure inflammatory lesion, consisting only of monocyte-derived macrophages and T lymphocytes (Stary et al. 1994).

1.2 Atherosclerosis and Macrophages

Cells of the monocyte-macrophage lineage are known to play a central role in atherogenesis (National Cholesterol Education Program, 1993). It has been known for many years that monocyte-derived macrophages are present in large numbers in atherosclerotic lesions (Faggiotto and Ross, 1984; Fowler et al. 1979; Gerrity, 1981; Schaffner et al. 1980).

In normal steady-state conditions, monoblasts and promonocytes are not released from the bone marrow into the peripheral blood unless they differentiate and mature first into monocytes (Takahashi, 2001; van Furth, 1989). After their release from the bone marrow, monocytes circulate in the peripheral blood and undergo apoptosis (Heidenreich, 1999; Kiener et al. 1997). In atherosclerotic lesions, many proinflammatory cytokines are produced by macrophages in response to infiltrated lipoproteins (Huber et al. 1999; Liu et al. 1997; Pinderski Oslund et al. 1999) and are released to activate circulating blood monocytes (Heidenreich, 1999). Oxidized low-density lipoprotein (Ox-LDL) stimulates monocytes to induce their migration into the arterial intima. In addition to Ox-LDL, many chemotactic factors for monocytes are known to be produced in atherosclerosis. The monocytes pass through the vascular endothelial cells and migrate into the subendothelial space of the intima. During these events, the vascular endothelial cells express various adhesion molecules essential for migration of monocytes.

As LDL circulates in blood or after it infiltrates the arterial wall, it is oxidized via enzymatic or nonenzymatic oxidation mechanisms (Heinecke, 1997). Ox-LDL is specifically taken up by macrophages via scavenger receptors (SRs) (Ross, 1993).

In atherosclerotic lesions of human aorta, the expression of CD36 is more marked in macrophage-derived foam cells than in native macrophages (Nakata et al. 1999).

Maintenance of cholesterol homeostasis is a challenge for all peripheral cells, and is particularly it is important for macrophages. In addition to the usual excess of free cholesterol resulting from intracellular membrane turnover, macrophages are exposed to large amounts of cholesterol derived from cell debris or from lipoproteins. The rapid internalization of this cholesterol cargo affects macrophage viability and function, and is an important element in the process of foam cell formation. Since the control of cholesterol efflux is of vital relevance for macrophages, it is possible that macrophage-specific mechanisms of cholesterol efflux may regulate lipid homeostasis efficiently in this cell type. Ox-LDL is specifically taken up by macrophages by scavenger receptors, which are responsible for removing deposited lipoproteins in the lesions. As scavenger receptor expression is not regulated by cellular cholesterol levels, Ox-LDL is continuously taken up, leading to cholesterol accumulation in macrophages and to foam cell formation. Ultimately, the cholesterol-loaded macrophages die and release their cholesterol content, causing pools of cholesterol to accumulate.

1.3 PON1 and Its Protective Role Against Atherosclerosis

Paraoxonase 1 (PON1) is a protein of 354 amino acids with a molecular mass of 43 kDa (Mackness et al. 1996; Primo-Parmo et al. 1996). Serum PON1 is a hydrolase which is synthesized by the liver and released into the serum, where it is associated with HDL. PON1 is present in a number of tissues such as the liver, kidney, heart, brain, small intestine, and lung (Primo-Parmo et al. 1996; Rodrigo et al. 2001), and was also found in atherosclerotic lesions (Aviram et al. 2000; Mackness et al. 1997). Elucidation of the crystal structure of PON1 revealed that PON1 is a six-bladed β-propeller, and each blade contains four strands (Harel et al. 2004).

PON1 was shown to exhibit a wide range of hydrolytic activities such as arylesterase, phosphotriesterase and lactonase with defined enzyme–lactone interaction (Tavori et al. 2008) (which best resemble PON1, physio/pathological functions (Gaidukov and Tawfik, 2005)).

Human serum PON1 activity was shown to be inversely related to the risk of cardiovascular diseases (Mackness et al. 2001), and low serum PON1 activities were observed in atherosclerotic, hypercholesterolemic and diabetic patients (Boemi et al. 2001), as well as in atherosclerotic apolipoprotein E-deficient (E^0) mice (Aviram et al. 1998), and in rabbits fed an atherogenic diet (Mackness et al. 2000). The role of PON1 in atherosclerosis development was demonstrated in studies using mice lacking PON1 (Shih et al. 1998; 2001), or overexpressing PON1 (Tward et al. 2002).

PON1 possesses anti-atherogenic properties including protection of LDL, HDL and macrophages against oxidative stress (Aviram and Rosenblat, 2004), attenuation of Ox-LDL uptake by macrophages (Fuhrman et al. 2002), inhibition of macrophage cholesterol biosynthesis (Rozenberg et al. 2003), and stimulation of HDL- mediated cholesterol efflux from the cells (Rosenblat et al. 2005). It was also shown that treatment of carotid lesions with PON1 results in reduction of the lipid oxidative potential (Tavori et al. 2008).

2 HDL

2.1 HDL Structure and Its Anti-atherogenic Properties

HDL particles consist of phospholipids, free cholesterol, cholesterol esters, and protein components, such as apolipoprotein A-I (apoA-I), A-II, Cs, and Es, as well as several enzymes such as LCAT, PAF-AH, and PON1 (Kakafika and Xenofontos, 1926; Silva et al. 1999) which was found to be associated with HDL through a direct binding of its N-terminal region to the HDL phospholipids (Sorenson et al. 1999).

HDL particles possess multiple anti-atherogenic activities, which include reverse cholesterol transport from the arterial wall to the liver for excretion, and anti-oxidative, anti-inflammatory, anti-apoptotic, anti-thrombotic, anti-infectious, and vasodilatory actions (Assmann and Nofer, 2003). Cholesterol efflux from peripheral blood cells, particularly from macrophages, constitutes the initial step of reverse cholesterol transport. Lipid-free HDL apolipoproteins, primarily apoA-I, cause specific efflux of cellular cholesterol and phospholipids via the ATP-binding cassette transporter A1 (ABCA1). In addition, HDL inhibits LDL oxidation, (Navab et al. 2004). Such activity was found to be functionally dependent on the presence of apolipoproteins and enzymes possessing anti-oxidative properties. ApoA-I, a major component of the anti-oxidative activity of HDL, can prevent LDL oxidation, by the removal of oxidized phospholipids from LDL, from artery wall cells, or from both. HDL has also cytokine-mediated expression of adhesion molecules, diminishes neutrophil infiltration within the arterial wall, and reduces generation of reactive oxygen species (Nicholls, 2005).

2.2 HDL Components Regulate PON1 Activity and Stability

HDL has been described as "an attractive vehicle from which PON1 could exert a protective, antioxidant function" (James and Deakin, 2004). The HDL component apoA-I was shown to stimulate PON1 catalytic activity (Deakin et al. 2002; Efrat et al. 2008; Oda et al. 2002). ApoA-I, a major HDL component, is also important for stabilizing the enzyme activity of the secreted protein and it was also shown that it protects PON1 from inactivation (Efrat et al. 2008; Kakafika and Xenofontos, 1926). Studies in human apoA-I deficiency diseases are consistent with this conclusion, as serum PON1 is reduced in such patients (James et al. 1998).

al.Several phospholipids were shown to stimulate and to stabilize PON1 activity (James et al. 1998; Kakafika and Xenofontos, 1926; La Du et al. 1993). Phosphatidylcholine (PC) makes up a very high proportion of the cell plasma membrane, and is the major phospholipid circulating in plasma, where it is an integral component of lipoproteins, especially of HDL (Small, 1992). PCs with saturated acyl chains increased PON1 paraoxonase and arylesterase activities, and this effect was shown to be acyl chain length-dependent (Aviram et al. 2000). Stimulation of PON1 arylesterase activity was remarkable with PCs containing polyunsaturated acyl chains or oxidized chains at the *sn*-2 position, in comparison to purified PON1 alone (Nguyen and Sok, 2006). PC with oleoyl fatty acids was shown to possess the most potent stimulatory effect on PON1 activities. Indeed, PON1 was found to be stabilized by oleic acid or by oleoylated phospholipids (Nguyen and Sok, 2003).

3 PON1, HDL, Dietary Antioxidants, and Their Interactions with Macrophages

3.1 Effects of HDL Phospholipids Composition on HDL and PON1 Interactions with Macrophages

Structural subclasses of HDL have been shown to exhibit varying abilities to accept cellular cholesterol (Castro and Fielding, 1988). Studies have determined that efficient cholesterol efflux requires the presence of the main protein component of HDL, apolipoprotein (apo) A-I, (Davidson et al. 1994) as well as that of phospholipids.

Modification of HDL phospholipids composition could be accomplished in two ways: via nutrition (in vivo by using food supplements) (James and Deakin, 2004; Kudchodkar et al. 2000; Rao et al. 1993; Ruiz-Gutierrez et al. 1990; Yaqoob et al. 1995) or by serum enrichment (James and Deakin, 2004; Jian et al. 1997).

The composition of HDL phospholipids play a major role, not only in HDL function (Davidson et al. 1994; Esteva et al. 1986; Jian et al. 1997), but also in regulating PON1 catalytic and biological activities in its interaction with macrophages. It was demonstrated (James and Deakin, 2004) both in vitro (using serum PON1/HDL PON1) and in vivo (in mice and in humans), that increasing HDL di-oleoyl phosphatidylcholine (PC-18:1) concentration significantly increases PON1 catalytic activities, and its ability to stimulate cholesterol efflux from macrophages by serum and HDL (Fig. 1). In vivo studies using dietary supplements of olive oil, which is rich in oleic acid (18:1), were performed. Thus, the increment in HDL PC-18:1 levels could have resulted from an exchange of the HDL phospholipids with serum phospholipids. The increased cholesterol efflux rate from macrophages by PC-18:1-enriched serum or HDL samples (in vitro and in vivo) could have resulted from the increment in phospholipids concentration, as phospholipids were shown to increase the capacity of HDL to promote cholesterol efflux from cultured cells, and this effect correlates well with the chain length of HDL phospholipids fatty acids, as well as with their saturation state (Esteva et al. 1986; Sola et al. 1993). In addition, it was clearly demonstrated that HDL PC-18:1 increased PON1 contribution to HDL-mediated macrophage cholesterol efflux.

Furthermore, it has been shown that the effect of PC-18:1 on PON1 is mediated, at least in part, via the PON1 N-terminal region (James and Deakin, 2004), the region which is responsible for PON1 binding to HDL (Kakafika and Xenofontos,

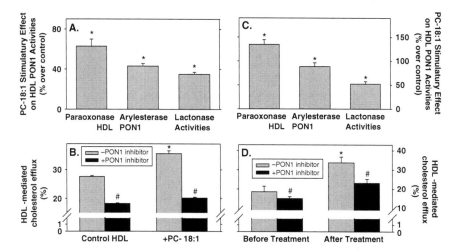

Fig. 1 The effects of HDL enrichment with PC-18:1 in vitro and in vivo on HDL-associated-PON1 catalytic and biological activities. (**A–B**): In vitro studies: Serum from healthy subjects was incubated with or without PC-18:1 (2.6 mM) for 2 h at 37°C. HDL was isolated from control serum or from PC-18:1-enriched serum by ultracentrifugation. (**a**) HDL-associated paraoxonase, arylesterase or lactonase activities were determined. (**b**) The HDL fractions were preincubated with 500 μM of PON1 specific inhibitor, 2-hydroxyquinoline. These HDL samples (100 μg protein/ml) were then further incubated with J774A.1 macrophages labeled with ^{3}[H]-cholesterol. The extent of HDL-mediated cholesterol efflux from macrophages was determined as described under the Methods section.

(**C-D**): In vivo studies: Healthy subjects (n=3) consumed olive oil (which is rich in oleic acid (18:1, 30 ml/day) for 2 weeks. Blood samples were collected before treatment and at the end of the study. The HDL fractions were isolated from these serum samples by ultracentrifugation. (**c**) HDL-associated paraoxonase, arylesterase or lactonase activities were determined. (**d**) J774A.1 macrophages were preincubated with ^{3}[H]-cholesterol, and the ability of the HDLs (100 μg protein/ml) to promote cholesterol efflux from the cells was measured as described under the Methods section. Results represent mean ± S.D of three different experiments. *$p < 0.01$ HDL PON1 vs. control serum, or *$p < 0.01$ after treatment vs. before treatment, #$p < 0.01$+PON1 inhibitor vs.–PON1 inhibitor

1926). Binding of PC-18:1 to PON1 N-terminal region could have changed PON1 conformation, and thus could possibly affect PON1 active site.

3.2 Binding and Internalization of HDL and PON1 into Macrophages

It was previously suggested by Sorenson et al. and James et al. that PON1 may act outside the HDL complex (Kakafika and Xenofontos, 1926; Nicholls, 2005) as it could transfer from HDL toward cell membranes within the artery wall, and exert its activity at the cell membrane region. James et al. (Nicholls, 2005) demonstrated that the transfer of PON1 from cell membrane to HDL provides support for a transfer process. PON1 had been demonstrated in atherosclerotic plaque areas, but not in healthy regions of the artery (Mackness et al. 1997), and an active enzyme has been extracted from such plaques (Aviram et al. 2000). Previous studies showed that under acute phase, where apoA-I, which stabilizes PON1, is displaced by serum amyloid A (SAA), HDL binding to hepatocytes was decreased whereas HDL binding to macrophages was dramatically increased (Kisilevsky and Subrahmanyan, 1992). This displacement may facilitate the release of PON1 from HDL and increases PON1 ability to react with macrophages because of the increase in HDL affinity to macrophages.

It was also shown (Rozenberg et al. 2003) that when MPM from PON1^0 mice was incubated with human PON1, a 40% increase in PON1 activity was noted. This phenomenon is in accordance with findings indicating PON1 location in the external membrane of PON1-transfected hepatocytes (Efrat et al. 2008).

A retroendocytosis of HDL was previously demonstrated in macrophages (Schmitz et al. 1985), where receptor-bound HDL internalizes into the cells. It was also suggested (Rozenberg et al. 2003) that through such a mechanism, PON1 can interact physiologically with macrophages, resulting in anti-atherogenic effects. Indeed, macrophage-bound PON1 is internalized and accumulates in the cell cytosolic compartment (Fig. 2) (Efrat and Aviram, 2008). It is of interest that PON1 accumulates in macrophages in the same compartment as PON2 protein (Shiner et al. 2007), which raises the possibility of an interaction between internalized exogenous PON1 and cellular endogenous PON2, thus allowing for better protection of macrophages from oxidative stress. These data could also be resolved with the hypothesis that PON1 interaction with macrophages will increase in the acute phase, via the displacement of apoA-I with SAA.

Both HDL and LDL specifically bind to macrophages (de Villiers and Smart, 1999; Schmitz and Schambeck, 2006). HDL, but not LDL, however, was able to

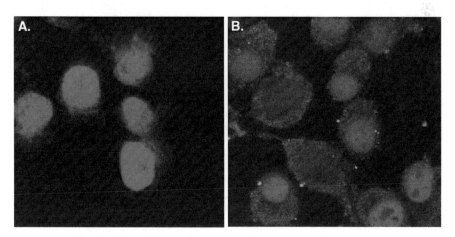

Fig. 2 PON1 association with macrophage cell membrane. Confocal laser scanning microscopic staining view of PON1 in J774A.1 macrophages following cell incubation at 37°C for 4 h with: (**a**) no addition, (**b**) addition of recombinant PON1 (0.4 mg/ml)

compete with PON1 for PON1 binding to macrophages (Schmitz et al. 1985), suggesting that an HDL component (PON1?) is responsible for HDL competition with labeled PON1 for PON1 binding sites on macrophages. In this respect, it should be noted that HDL, but not LDL, was also shown to be able to facilitate PON1 secretion from cells (Nicholls, 2005).

Binding assays (Schmitz et al. 1985) showed that PON1 binding to macrophages reached saturation, but with relatively low specificity, suggesting a limited number of receptors for PON1 on macrophages (Schmitz et al. 1985). A possible mechanism could be HDL binding via macrophage SR-B1 that anchors HDL to the cell membrane in order to allow for an exchange of PON1 from the lipoprotein to the cell surface. In this respect, Tavori et al. demonstrated that inhibition of PON1 lactonase activity correlates with inhibition of HDL ability to promote efflux from macrophages (Tavori et al. 2008).

In addition, Rosenbalt et al. (Rosenblat et al. 2005) showed in their studies that PON1 affects HDL interaction with macrophages. HDL PON1 concentration is directly correlated with the extent of HDL binding to macrophages and to HDL ability to promote cholesterol efflux from macrophages (Rosenblat et al. 2005). The above authors suggested a mechanism whereby the HDL-associated PON1 hydrolytic action on macrophage phospholipids induces the formation of LPC, which then increases HDL binding to the cells; this effect can contribute to the observed enhanced macrophage cholesterol efflux. The ability of PON1 to stimulate HDL-mediated macrophage cholesterol efflux was shown following human HDL enrichment with purified PON1, as well as by using HDL from PON1[0] mice or from PON1Tg mice. This stimulatory effect of PON1 on HDL binding to the cells could have resulted from PON1-induced conformational changes in the lipoprotein particle, as well as from PON1 increased binding to cellular specific phospholipids, thus increasing the binding of HDL to its receptor (Rosenblat et al. 2005). By measuring cellular cholesterol mass, it was demonstrated that the effect of HDL-associated PON1 was indeed on net cholesterol efflux and not the result of a cholesterol exchange phenomenon (Rosenblat et al. 2005).

Figure 3 summarizes the complex relationship between PON1, dietary antioxidants, HDL, and macrophages.

3.3 A Possible Role for Phospholipids in HDL PON1–Macrophage Interactions

Secreted PON1 retains its hydrophobic N-terminal sequence, which enables its association with phospholipids, on lipoproteins and on cells (Gaidukov and Tawfik, 2005; Kakafika and Xenofontos, 1926; Schmitz et al. 1985). PON1's retained N-terminal peptide may allows for the transfer of PON1 between phospholipid surfaces. It is possible that PON1 binds to the macrophage's plasma membrane via cellular phospholipids and then enters the cell, via a "flip-flop" transport mechanism (Schmitz et al. 1985).

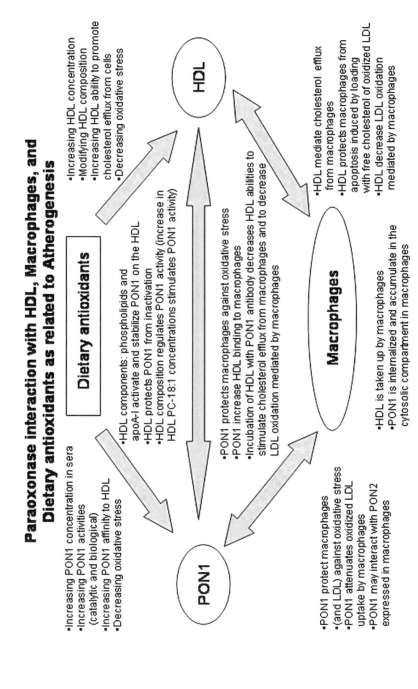

Fig. 3 PON1 interactions with HDL, macrophages, and dietary antioxidants as related to atherogenesis

4 Future Perspectives

Macrophage cholesterol accumulation and foam cells are the hallmark of early atherogenesis (Aviram, 1993; Witztum and Steinberg, 1991). Cholesterol accumulation in these cells can result from increased uptake of oxidized LDL, increased cholesterol biosynthesis rate, and/or decreased rate of HDL-mediated cholesterol efflux from the cells (Aviram, 1996; Krieger, 1998; Steinberg, 1997).

The present research demonstrated that modifying HDL phospholipid composition can regulate HDL and PON1 interactions with macrophages. The increase shown in macrophage cholesterol efflux by PC-18:1-enriched HDL, through PON1 activation, can be the main contributor to attenuation of atherosclerosis development. However, PON1 protects against macrophage-mediated LDL oxidation, and also increases HDL binding to macrophages which, in turn, stimulates HDL ability to promote cholesterol efflux (Fuhrman et al. 2002; Rosenblat et al. 2005). These two major anti- atherogenic properties of HDL (and of PON1), require, at least in part, macrophages binding sites for the HDL-associated PON1. PON1, as well as HDL-associated PON1, specifically binds to macrophages, leading to anti-atherogenic effects. Macrophage PON1 binding sites may thus be a target for future cardioprotection therapy. Thus, studying the interactions between PON1 and macrophages can help in achieving appropriate treatment for atherosclerosis. There are several approaches that could assist in understanding these interactions: increasing PON1 concentration, increasing HDL binding to macrophages, increasing PON1 binding sites on macrophages, and also modifying the phospholipid composition of HDL and cell membranes.

References

Assmann G and Nofer JR (2003) Atheroprotective effects of high-density lipoproteins. Annu Rev Med 54:321–341

Aviram M. (1993) Modified forms of low density lipoprotein and atherosclerosis. Atherosclerosis98:1–9.

Aviram M (1996) Interaction of oxidized low density lipoprotein with macrophages in atherosclerosis, and the antiatherogenicity of antioxidants. Eur J Clin Chem Clin Biochem 34:599–608.

Aviram M, Hardak E, Vaya J et al. (2000) Human serum paraoxonases (PON1) Q and R selectively decrease lipid peroxides in human coronary and carotid atherosclerotic lesions: PON1 esterase and peroxidase-like activities. Circulation 101:2510–2517.

Aviram M, Rosenblat M (2004) Paraoxonases 1, 2, and 3, oxidative stress, and macrophage foam cell formation during atherosclerosis development. Free Radic Biol Med 37:1304–1316.

Aviram M, Rosenblat M, Bisgaier C et al. (1998) Paraoxonase inhibits high-density lipoprotein oxidation and preserves its functions. A possible peroxidative role for Paraoxonase. J Clin Invest 101:1581–1590.

Boemi M, Leviev I, Sirolla C et al. (2001) Serum paraoxonase is reduced in type 1 diabetic patients compared to non-diabetic, first degree relatives: influence on the ability of HDL to protect LDL from oxidation. Atherosclerosis 155:229–235

Braunwald EShattuck Lecture (1997) Cardiovascular medicine at the turn of the millennium: triumphs, concerns, and opportunities. N Engl J Med 337:1360–1369.

Breslow JL (1997) Cardiovascular disease burden increases, NIH funding decreases. Nat Med 3:600–601.

Castro GR, Fielding CJ. (1988) Early incorporation of cell derived cholesterol into pre-beta-migrating high-density lipoprotein. Biochemistry 27: 25–29.

Davidson WS, Lund-Katz S, Johnson WJ et al. (1994) The influence of apolipoprotein structure on the efflux of cellular free cholesterol to high density lipoprotein. J Biol Chem 269: 22975–22982.

W.J. de Villiers, E.J. Smart (1999) Macrophage scavenger receptors and foam cell formation J Leukoc Biol 66:740–746.

Deakin S, Leviev I, Gomaraschi M et al.(2002) Enzymatically active paraoxonase-1 is located at the external membrane of producing cells and released by a high affinity, saturable, desorption mechanism. J Biol Chem 277:4301–4308.

Efrat M, Aviram M (2008) Macrophage paraoxonase 1 (PON1) binding sites. Biochem Biophys Res Commun 376:105–110.

Efrat M, Rosenblat M, Mahmood S et al. (2008) Di-oleoyl phosphatidylcholine (PC-18:1) stimulates paraoxonase 1 (PON1) enzymatic and biological activities: In vitro and in vivo studies. Atherosclerosis. [Epub ahead of print]

Esteva O, Baudet MF, Lasserre M et al. (1986) Influence of the fatty acid composition of high-density lipoprotein phospholipids on the cholesterol efflux from cultured fibroblasts. Biochim Biophys Acta 875:174–182.

Faggiotto A, Ross R. (1984) Studies of hypercholesterolemia in the nonhuman primate, I: changes that lead to fatty streak formation. Arteriosclerosis 4:323–340.

Fowler S, Shio H, Haley NJ (1979) Characterization of lipid-laden aortic cells from cholesterol-fed rabbits, IV: investigation of macrophage-like properties of aortic cell populations. Lab Invest 41:372–378.

Fuhrman B, Volkova N, Aviram M. (2002) Oxidative stress increases the expression of the CD36 scavenger receptor and the cellular uptake of oxidized low-density lipoprotein in macrophages from atherosclerotic mice: protective role of antioxidants and of paraoxonase. Atherosclarosis161:307–316.

31Gaidukov L, Tawfik D (2005) High affinity, stability, and lactonase activity of serum paraoxonase PON1 anchored on HDL with ApoA-I. Biochemistry. 44:11843–11854.

Gerrity RG (1981) The role of the monocyte in atherogenesis, II: migration of foam cells from atherosclerotic lesions. Am J Pathol 103:191–200

Harel M, Aharoni A, Gaidukov L et al. (2004) Structure and evolution of the serum paraoxonase family of detoxifying and anti-atherosclerotic enzymes. Nat Struct Mol Biol 11:412–419.

Heidenreich S(1999). Monocyte CD14: a multifunctional receptor engaged in apoptosis from both sides. J Leukoc Biol 65:737–743

Heinecke J (1997) Mechanisms of oxidative damage of low density lipoprotein in human atherosclerosis. Curr Opin Lipido. 18:268–274

Huber SA, Sakkinen P, Conze D et al. (1999) Interleukin-6 exacerbates early atherosclerosis in mice. Arterioscler Thromb Vasc Biol 19:2364–2367

Idem M (1976) The pathogenesis of atherosclerosis. N Engl J Med 295:369–377, 420–425.

James, R, Blatter Garin M, Calabresi L et al. (1998) Modulated serum activities and concentrations of paraoxonase in high density lipoprotein deficiency states. Atherosclerosis. 139:77–82

James R, Deakin S (2004) The importance of high-density lipoproteins for paraoxonase-1 secretion, stability, and activity. Free Radic Biol Med 37:1986–1994

Jian B, de la Llera-Moya M, Royer L et al. (1997) Modification of the cholesterol efflux properties of human serum by enrichment with phospholipid. J Lipid Res 38:734–744

Kakafika AI, Xenofontos S, Tsimihodimos V (2003) The PON1 M55L gene polymorphism is associated with reduced HDL-associated PAF-AH activity. J Lipid Res 44:1919–1926.

Kiener PA, Davies PM, Starling GC et al. (1997) Differential induction of apoptosis by Fas-Fas ligand interaction in human monocytes and macrophages. J Exp Med 185:1511–1516

Kisilevsky R, Subrahmanyan L (1992) Serum amyloid A changes high density lipoprotein's cellular affinity. A clue to serum amyloid A's principal function. Lab Invest 66: 778–785

Krieger M (1998) The "best" of cholesterols, the "worst" of cholesterols: a tale of two receptors. Proc Natl Acad Sci USA 14:4077–4080.

Kudchodkar B, Lacko A, Dory L, Fungwe T (2000) Dietary fat modulates serum paraoxonase 1 activity in rats. Nutr 130:2427–2433

La Du B, Adkins S, Kuo C et al. (1993) Studies on human serum paraoxonase/ arylesterase. Chem Biol Interact 87:25–34.

Liu Y, Hulten LM, Wiklund O (1997) Macrophages isolated from human atherosclerotic plaques produce IL-8, and oxysterols may have a regulatory function for IL-8 production. Arterioscler Thromb Vasc Biol 17:317–323

Mackness M, Boullier H, Hennuyer M et al. (2000) Paraoxonase activity is reduced by a proatherogenic diet in rabbits. Biochem Biophys Res Commun 269:232–236.

Mackness B, Davies G, Turkie W et al. (2001) Paraoxonase status in coronary heart disease: are activity and concentration more important than genotype? Arterioscler Thromb Vasc Biol 211:451–457

Mackness B, Hunt R, Durrington PN (1997) Increased immunolocalization of paraoxonase, clusterin, and apolipoprotein A-I in the human artery wall with the progression of atherosclerosis. Arterioscler Thromb Vasc Biol 17:1233–1238.

Mackness M, Mackness B, Durrington P et al. (1996) Paraoxonase: biochemistry, genetics and relationship to plasma lipoproteins. Curr Opin Lipidol 7:69–76.

Nakata A, Nakagawa Y, Nishida M et al. (1999) CD36, a novel receptor for oxidized low-density lipoproteins, is highly expressed on lipidladen macrophages in human atherosclerotic aorta. Arterioscler Thromb Vasc Biol 19:1333–1339.

Napoli C, D'Armiento F, Mancini F et al. (1997) Fatty streak formation occurs in human fetal aortas and is greatly enhanced by maternal hypercholesterolemia: intimal accumulation of low density lipoprotein and its oxidation precede monocyte recruitment into early atherosclerotic lesions. J Clin Invest 100:2680–2690.

National Cholesterol Education Program. (1993) Second report of the Expert Panel on Detection, Evaluation, and Treatment of High Blood Cholesterol in Adults (Adult Treatment Panel II). Bethesda, Md.: National Heart, Lung, and Blood Institute. (NIH publication no. 93-3095.)

Navab M, Ananthramaiah GM, Reddy ST et al. (2004) The oxidation hypothesis of atherogenesis: the role of oxidized phospholipids and HDL. J Lipid Res 45: 993–1007

Nguyen S, Sok D(2003) Beneficial effect of oleoylated lipids on paraoxonase 1: protection against oxidative inactivation and stabilization. Biochem J 15375:275–285.

Nguyen SD, Sok DE (2006) Preferable stimulation of PON1 arylesterase activity by phosphatidylcholines with unsaturated acyl chains or oxidized acyl chains at sn-2 position. Biochim Biophys Acta 1758:499–508.

Nicholls SJ et al. (2005) Reconstituted high-density lipoproteins inhibit the acute pro-oxidant and proinflammatory vascular changes induced by a periarterial collar in normocholesterolemic rabbits. Circulation 111: 1543–1550

Oda, M, Bielicki J, Berger T et al. (2002) Cysteine substitutions in apolipoprotein A-I primary structure modulate paraoxonase activity. Biochemistry 40:1710–1718

Pinderski Oslund L, Hedrick C, Hagenbaugh A et al. (1999) Interleukin-10 blocks atherosclerosis events in vivo and in vitro. Arterioscler Thromb Vasc Biol 19:2847–2853

Primo-Parmo SL, Sorenson RC, Teiber J, La Du BN (1996) The human serum paraoxonase/arylesterase gene (PON1) is one member of a multigene family. Genomics 33:498–507.

Rao C, Zang E, Reddy B (1993) Effect of high fat corn oil, olive oil and fish oil on phospholipid fatty acid composition in male F344 rats. Lipids 28:441–447

Rodrigo L, Hernandez A, Lopez-Caballero J (2001) Immunohistochemical evidence for the expression and induction of paraoxonase in rat liver, kidney, lung and brain tissue. Implications for its physiological role. Chem Biol Interact 137:123–137.

Rosenblat M, Vaya J, Shih D et al.(2005) Paraoxonase 1 (PON1) enhances HDL-mediated macrophage cholesterol efflux via the ABCA1 transporter in association with increased HDL binding to the cells: a possible role for lysophosphatidylcholine. Atherosclarosis. 179:69–77.

Ross R (1986) The pathogenesis of atherosclerosis — an update. N Engl J Med 314:488–500.

Ross R (1993) Atherosclerosis: defense mechanisms gone awry. Am J Pathol 143:987–1002

Ross R, Glomset J (1973) Atherosclerosis and the arterial smooth muscle cell: proliferation of smooth muscle is a key event in the genesis of the lesions of atherosclerosis. Science 180: 1332–1339.

Rozenberg O, Shih DM, Aviram M (2003) Human serum paraoxonase 1 decreases macrophage cholesterol biosynthesis: possible role for its phospholipase-A2-like activity and lysophosphatidylcholine formation. Arterioscler Thromb Vasc Biol 23:461–467.

Ruiz-Gutierrez V, Molina M, Vazquez C (1990) Comparative effects of feeding different fats on fatty acid composition of major individual phospholipids of rat hearts. Ann Nutr Metab 34: 350–358

Scandinavian Simvastatin Survival Study Group (1994) Randomised trial of cholesterol lowering in 4444 patients with coronary heart disease: the Scandinavian Simvastatin Survival Study (4S). Lancet 344:1383–1389.

Schaffner T, Taylor K, Bartucci E et al. (1980) Arterial foam cells with distinctive immunomorphologic and histochemical features of macrophages. Am J Pathol 100:57–80.

Schmitz G, Robenek H, Lohmann U, Assmann G(1985) Interaction of HDLs with cholesteryl ester-laden macrophages: biochemical and morphological characterization of cell surface receptor binding, endocytosis and resecretion of HDLs by macrophages. EMBO J 4:613–622.

Schmitz G, Schambeck C (2006) Molecular defects in the ABCA1 pathway affect platelet function. Pathophysiol Haemost Thromb 35:166–174.

Shih D, Gu L, Xia Y et al (1998) Mice lacking serum paraoxonase are susceptible to organophosphate toxicity and atherosclerosis. Nature 394:284–287.

Shih D, Xia Y, Miller E et al.(2001) Combined serum paraoxonase knockout/apolipoprotein E knockout mice exhibit increased lipoprotein oxidation and atherosclerosis. J. Biol. Chem 275:7527–17535

Shiner M, Fuhrman B, Aviram M (2007) Macrophage paraoxonase 2 (PON2) expression is upregulated by unesterified cholesterol through activation of the phosphatidylinositol 3-kinase (PI3K) pathway. Biol Chem 388:1353–1358.

Silva E, Kong M, Han Z (1999) Metabolic and genetic determinants of HDL metabolism and hepatic lipase activity in normolipidemic females. J Lipid Res 40:1211–1221.

Small DM (1992) Plasma lipoprotein and coronary artery disease, (Kreisberg R.A, Segrest J.P (eds.), 1st ed., Boston: Blackwell scientific publications, chapter 3, 57–84

Sola R, Motta C, Maille M et al. (1993) Dietary monounsaturated fatty acids enhance cholesterol efflux from human fibroblasts. Relation to fluidity, phospholipid fatty acid composition, overall composition, and size of HDL3. Arterioscler Thromb 13:958–966.

Sorenson RC, Bisgaier CL, Aviram M (1999) Human serum paraoxonase/ arylesterase's retained hydrophobic N-terminal leader sequence associates with HDLs by binding phospholipids; apolipoprotein A 1 stabilizes activity. Arterioscler Thromb Vasc Biol 19:2214–2225.

Stary H, Chandler A, Glagov S (1994) A definition of initial, fatty streak, and intermediate lesions of atherosclerosis: a report from the Committee on Vascular Lesions of the Council on Arteriosclerosis, American Heart Association. Circulation 89:2462–2478.

Steinberg D (1997) Low density lipoprotein oxidation and its pathobiological significance. J Biol Chem 272:20963–20966.

Takahashi K (2001) Development and differentiation of macrophages and related cells: historic review and current concepts. J Clin Exp Hematopathol 41:1–32

Tavori H, Aviram, M, Khatib S et al. (2008) Human carotid atherosclerotic plaque increases oxidative stress of macrophages and LDL, whereas Paraoxonase 1 (PON1) decreases such atherogenic effects. Free Radic Biol Med [In Review]

Tavori H, Khatib S, Aviram M et al. (2008) Characterization of the PON1 active site using modeling simulation, in relation to PON1 lactonase activity. Bioorg Med Chem 16:7504–7509.

Tward A, Xia Y, Wang X, (2002) Decreased atherosclerotic lesion formation in human serum paraoxonase transgenic mice. Circulation 106:484–490 van Furth R. (1989) Origin and turnover of monocytes and macrophages. In: Iverson O (ed) Cell kinetics of inflammatory reaction. Berlin: Springer 125–150

Witztum J, Steinberg, D (1991) Role of oxidized low density lipoprotein in atherogenesis Clin Invest 88:1785–1792.

Yaqoob P, Sherrington E, Jeffery N et al. (1995) Comparison of the effects of a range of dietary lipids upon serum and tissue lipid composition in the rat. Int J Biochem Cell Biol 27:297–310

The Effect of HDL Mimetic Peptide 4F on PON1

Ladan Vakili, Susan Hama, J. Brian Kim, Duc Tien, Shila Safarpoor, Nancy Ly, Ghazal Vakili, Greg Hough, and Mohamad Navab

Abstract Several lines of evidence indicate that serum paraoxonase 1 (PON1) acts as an important guardian against cellular damage from oxidized lipids in plasma membrane, in low-density lipoprotein (LDL), against bacterial endotoxin and against toxic agents such as pesticide residues including organophosphates. In circulation, the high-density lipoprotein (HDL)-associated PON1 has the ability to prevent the formation of proinflammatory oxidized phospholipids. These oxidized phospholipids negatively regulate the activities of the HDL-associated PON1 and several other anti-inflammatory factors in HDL. During the acute phase response in rabbits, mice, and humans, there appears to be an increase in the formation of these oxidized lipids that results in the inhibition of the HDL-associated PON1 and an association of acute phase proteins with HDL that renders HDL proinflammatory. Low serum HDL is a risk factor for atherosclerosis and attempts are directed toward therapies to improve the quality and the relative concentrations of LDL and HDL. Apolipoprotein A-I (apoA-I) has been shown to reduce atherosclerotic lesions in laboratory animals. ApoA-I, however, is a large protein and needs to be administered parenterally, and it is costly. We have developed apoA-I mimetic peptides that are much smaller than apoA-I, and much more effective in removing the oxidized phospholipids and other oxidized lipids. These mimetic peptides improve LDL and HDL composition and function and reduce lesion formation in animal models of atherogenesis. Following is a brief description of some of the HDL mimetic peptides that can improve HDL and the effect of the peptide on PON1 activity.

Keywords HDL · ApoA-I · Inflammation · D-4F · L-4F · Dysfunctional HDL · Proinflammatory HDL · Anti-inflammatory

1 Can Apolipoprotein A-I Serve as a Therapeutic Agent?

Preliminary human studies (Nissen et al., 2003) and experiments in animal models of atherosclerosis (Badimon et al., 1990; Plump et al., 1994) have made

M. Navab (✉)
David Geffen School of Medicine, University of California, Los Angeles, CA, USA
e-mail: mnavab@mednet.ucla.edu

S.T. Reddy (ed.), *Paraoxonases in Inflammation, Infection, and Toxicology*, Advances in Experimental Medicine and Biology 660, DOI 10.1007/978-1-60761-350-3_15, © Humana Press, a part of Springer Science+Business Media, LLC 2010

apolipoprotein A-I (apoA-I), the main protein in high-density lipoprotein (HDL), an attractive therapeutic target. The preliminary human studies (Nissen et al., 2003) suggested that therapeutic benefit might be achieved by administering weekly intravenous doses over a period of 5–6 weeks. Subsequent larger clinical trials (Tardif et al. 2007), however, suggested that for significant improvements to be achieved, longer periods of intravenous administration will be required, making this an unlikely therapy for the large number of patients with atherosclerosis.

2 ApoA-I Mimetic Peptide, the Search for an Ideal Agent

Originally, an 18 amino acid peptide was designed that did not have sequence homology with apoA-I but mimicked the class A amphipathic helixes contained in apoA-I (Anantharamaiah et al., 1985; Venkatachalapathi et al., 1993; Yancey et al., 1995). This peptide was called 18A because it contained 18 amino acids and formed a class A amphipathic helix. The stability and lipid-binding properties of 18A peptide were improved by blocking the amino terminus through the addition of an acetyl group and blocking the carboxy terminus using an amide group. This peptide was named 2F because of the presence of the two phenylalanine residues on the hydrophobic face. The 2F peptide mimicked many of the lipid-binding properties of apoA-I but failed to alter lesions in a mouse model of atherosclerosis (Datta et al., 2001). With the use of a human artery wall cell culture assay, we tested a series of peptides for their ability to inhibit low-density lipoprotein (LDL)-induced monocyte chemotactic activity, which is primarily due to the production of monocyte chemoattractant-1 (MCP-1). Subsequently, the peptide 4F, with the same structure as 2F except for two additional phenylalanine residues on the hydrophobic face of the peptide (replacing two leucine residues), was shown to be superior (Datta et al., 2001).

3 The Effects of Apolipoprotein Mimetic Peptides in Animal Models of Atherosclerosis

The development of atherosclerotic lesions in young mice was reduced by apoA-I mimetic peptides (Navab et al., 2002; Li et al., 2004) and, when 4F was given together with a statin, it caused lesion regression in old apolipoprotein E null mice (Navab et al., 2005a). In these studies we tested for synergy between D-4F (made from all D amino acids) and pravastatin. We first determined the oral doses for each that were ineffective when given as single agent. The oral administration of the combination significantly increased HDL cholesterol levels, apoA-I levels, and paraoxonase (PON1) activity. It additionally rendered HDL anti-inflammatory, significantly prevented lesion formation in young apolipoprotein E null mice, and caused regression of established lesions in old apolipoprotein E null mice. The mice that received the combination for 6 months had lesion areas that were smaller when compared to those before the start of treatment (Navab et al., 2005a). En face lesion area was 38% of that in mice maintained on chow alone after 6 months of treatment with the combination. In addition there was a significant 22% reduction in macrophage content in the remaining lesions, indicating an overall reduction in macrophages of 79%. The combination of statin and D-4F increased intestinal apoA-I synthesis by a significant 60%. In studies in Cynomolgous monkeys also, HDL was rendered anti-inflammatory by oral administration of D-4F plus pravastatin (Navab et al., 2005a). These results suggested that the combination of an HDL-based therapy and a statin might be a useful anti-atherosclerosis treatment strategy. We subsequently found that the benefit of apolipoprotein mimetic peptides in atherosclerosis was not limited to 4F. We showed that other mimetic peptides including D-[113–122]apoJ, an apolipoprotein J mimetic peptide (Navab et al., 2005b), and peptides too small to form a helical structure (Navab et al., 2005c) were also efficacious. The efficacy of these peptides was recently also demonstrated in a rabbit model of atherosclerosis (Van Lenten et al., 2007).

4 What is the Mechanism of Action of Apolipoprotein Mimetic Peptides?

The mechanism of action of the apolipoprotein mimetic peptides appears to be related to their ability to remove oxidized lipids from lipoproteins (Navab et al., 2004; 2005d; Datta et al. 2004), promote reverse cholesterol transport from macrophages, and render HDL anti-inflammatory (Navab et al., 2004; 2005d). Atherosclerosis is a long-term inflammatory process that is mediated in part by the oxidation of phospholipids, which induce vascular cells to express cytokines, adhesion molecules, and procoagulant molecules (Berliner and Watson, 2005; Gargalovic et al., 2006). The mechanism of action of the apolipoprotein mimetic peptides seems to be related to their ability to bind and remove these proinflammatory oxidized lipids (Navab et al., 2004; 2005d). In addition, the apolipoprotein mimetic peptides are efficacious in models of vascular diseases that are not classified as atherosclerosis and in inflammatory processes that have an infectious etiology, suggesting that oxidized lipids may be important mediators of a variety of inflammatory conditions other than atherosclerosis.

5 Paraoxonase, Oxidized Lipids and Apolipoprotein Mimetic Peptides

We originally reported that on an atherogenic diet, PON1 activity decreased by 52% in fatty streak susceptible mice, C57BL/6 J (BL/6), but not in fatty streak resistant mice, C3H/HeJ (C3H) (Navab et al. 1997). Plasma PON1 activity was also significantly decreased in apolipoprotein E knockout mice on the chow diet, as compared to controls. Furthermore, a significant decrease in PON activity was observed in LDL receptor knockout mice when they were fed a 0.15%-cholesterol-enriched diet. Injection of mildly oxidized LDL but not native LDL into BL/6 mice but not in C3H mice on a chow diet resulted in a 59% decrease in PON activity. Treatment of HepG2 cells in culture with mildly oxidized LDL (but not native LDL) resulted in a three-fold reduction in mRNA levels for PON1. We additionally reported that in normolipidemic patients with angiographically documented coronary artery disease who did not have diabetes and were not on lipid-lowering medication, the total cholesterol/HDL cholesterol ratio was 3.1±0.9 as compared to 2.9±0.4 in the controls (Navab et al., 1997). This difference was not statistically significant. In a subset of these normolipidemic patients, the PON1 activity was low (48±6.6 versus 98±17 U/ml for controls; $p = 0.009$), despite similar normal HDL levels, and the HDL from these patients failed to protect against LDL oxidation in co-cultures of human artery wall cells.

Forte and colleagues showed (Forte, 2002) that ApoE –/–mice had elevated lysophosphatidylcholine and bioactive oxidized phospholipids (1-palmitoyl-2-oxovaleryl-sn-glycero-3-phosphocholine and 1-palmitoyl-2-glutaryl-sn-glycero-3-phosphocholine) compared with controls on the chow diet. Elevated oxidized phospholipids may, in part, contribute to spontaneous lesions in these mice on a chow diet. A Western diet decreased PON1 activity in these mice by 38%. It has been suggested that the removal of oxidized fatty acids from HDL might cause the return of PON1 activity.

We have observed that when apoE-deficient mouse plasma is incubated with PON1, followed by FPLC fractionation of lipoproteins, PON1 is found in addition to in HDL-containing fractions, also in the post HDL region. We have shown that early during the incubation, PON is present in HDL-like particles that co-elute with albumin, and contain cholesterol, apoA-I, and other HDL constituents. It is possible that 4F accelerates apoA-I cycling, pre-beta formation and remodeling of HDL.

We have also observed that when human plasma is incubated with PON1 in vitro, PON1 activity is increased. As for the underlying mechanism, one possibility could be the removal of oxidized lipids from HDL and reactivation of PON1 by 4F. Another might be the changes in lipid–protein interaction such as that in phospholipid–apoA-I–PON1 interactions. 4F therefore might have beneficial effects removing oxidized lipids, reactivating antioxidant enzymes including PON1, and supporting HDL function in individuals under conditions that result in low PON1 activity.

6 Summary

Although PON1 has the ability to prevent lipid oxidation and may even inactivate oxidized lipids once formed, and thus protect HDL against the inflammatory pressure, under conditions of excess inflammatory pressure the ability of HDL to protect itself and other lipid containing molecules and structures might be reduced, HDL be damaged, antioxidant enzymes such as PON1 be inactivated, and even HDL

Fig. 1 Effect of 4F on PON1 distribution in mouse plasma. Fasting ApoE deficient mice were administered with scrambedd 4F (50 ug per mouse, *top panel*) or with active 4F (50 ug per mouse, *lower panel*). Four hours later blood was obtained and plasm a fractionated using FPLC, followed by cholesterol and PON1 activity determination

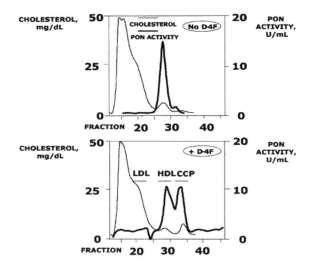

itself act as a proinflammatory molecule. Reduction of lipid and protein oxidation by agents such as HDL mimetic peptides may prove to be an effective way of supporting the protective role of HDL.

References

Anantharamaiah GM, Jones JL, Brouillette CG, et al. 1985. Studies of synthetic peptide analogs of amphipathic helix I: structure of peptide/DMPC complexes. *J Biol Chem* 260:10248–10255.

Badimon JA, Badimon L, Fuster V 1990. Regression of atherosclerotic lesions by high density lipoprotein plasma fraction in the cholesterol-fed rabbit. *J Clin Invest* 85:1234–1241.

Berliner JA, Watson AD 2005. A role for oxidized phospholipids in atherosclerosis. *N Engl J Med* 353:9–12.

Datta G, Chaddha M, Hama S, et al. 2001. Effects of increasing hydrophobicity on the physical–chemical and biological properties of a class A amphipathic helical peptide. *J Lipid Res* 42:1096–1104.

Datta G, Epand RF, Epand RM, et al. 2004. Aromatic residue position on the nonpolar face of class a amphipathic helical peptides determines biological activity. *J Biol Chem.* 18;279: 26509–26517.

Forte, TM, et al. 2002. Altered activities of anti-atherogenic enzymes LCAT, paraoxonase, and platelet-activating factor acetylhydrolase in atherosclerosis susceptible mice. *J Lipid Res* 43:477–485.

Gargalovic PS, Imura M, Zhang B, et al. 2006. Identification of inflammatory gene modules based on variations of human endothelial cell responses to oxidized lipids. *Proc Natl Acad Sci* 103:12741–12746.

Li X, Chyu K-Y, Faria JR, et al. 2004. Differential effects of apolipoprotein A-I mimetic peptide on evolving and established atherosclerosis in apolipoprotein E-null mice. *Circulation* 110: 1701–1705.

Navab M, Anantharamaiah GM, Hama S, et al. 2002. Oral administration of an apoA-I mimetic peptide synthesized from D-amino acids dramatically reduces atherosclerosis in mice independent of plasma cholesterol. *Circulation* 105:290–292.

Navab M, Anantharamaiah GM, Reddy ST, et al. 2004. Oral D-4F causes formation of pre-β high-density lipoprotein and improves high-density lipoprotein-mediated cholesterol efflux and reverse cholesterol transport from macrophages in apolipoprotein E-null mice. *Circulation* 109:3215–3220.

Navab M, Anantharamaiah GM, Hama S, et al. 2005a. D-4F and statins synergize to render HDL anti-inflammatory in mice and monkeys and cause lesion regression in old apolipoprotein E-null mice. *Arterioscler Thromb Vasc Biol* 25:1426–1432.

Navab M, Anantharamaiah GM, Reddy ST, et al. 2005b. An oral apoJ peptide renders HDL anti-inflammatory in mice and monkeys and dramatically reduces atherosclerosis in apolipoprotein E-null mice. *Arterioscler Thromb Vasc Biol* 25:1932–1937.

Navab M, Anantharamaiah GM, Reddy ST, et al. 2005c. Oral small peptides render HDL anti-inflammatory in mice and monkeys and reduce atherosclerosis in apoE null mice. *Circ Res* 97:524–532.

Navab M, Anantharamaiah GM, Reddy ST, et al. 2005d. Apolipoprotein A-I mimetic peptides. *Arterioscler Thromb Vasc Biol* 25:1325–1331.

Navab M, Hama-Levy S, Van Lenten BJ, et al. 1997. Mildly oxidized LDL induces an increased apolipoprotein J/paraoxonase ratio. *J Clin Invest.* 15;99:2005–2019.

Nissen SE, Tsunoda T, Tuzcu EM, et al. 2003. Effect of recombinant apoA-I Milano on coronary atherosclerosis in patients with acute coronary syndromes: A randomized controlled trial. *JAMA* 290:2292–2300.

Plump AS, Scott CJ, Breslow JL 1994. Human apolipoprotein A-I gene expression increases high density lipoprotein and suppresses atherosclerosis in the apolipoprotein E-deficient mouse. *Proc Natl Acad Sci U S A* 91:9607–9611.

Tardif JC, Grégoire J, L'Allier PL, et al. 2007. Effect of rHDL on atherosclerosis-safety and efficacy (ERASE) investigators. *JAMA.* 18;297:1675–1682.

Van Lenten BJ, Wagner AC, Navab M, et al. 2007. Lipoprotein inflammatory properties and serum amyloid A levels but not cholesterol levels predict lesion area in cholesterol-fed rabbits. *J Lipid Res* 48:2344–2353.

Venkatachalapathi YV, Phillips MC, Epand RM, et al. 1993. Effect of end group blockage on the properties of a class A amphipathic helical peptide. *Proteins Struct Funct Genet* 15:349–359.

Yancey PG, Bielicki JK, Lund-Katz S, et al. 1995. Efflux of cellular cholesterol and phospholipid to lipid-free apolipoproteins and class A amphipathic peptides. *Biochemistry* 34:7955–7965. PII S1050-1738(08)00002-9 TCM

The Contribution of High Density Lipoprotein Apolipoproteins and Derivatives to Serum Paraoxonase-1 Activity and Function

Richard W. James and Sara P. Deakin

Abstract High density lipoproteins (HDL) not only provide a serum transport vector for paraoxonase-1 (PON1) but also contribute to enzyme activity, stability and, consequently, function. The contribution of the apolipoprotein (apo) components of HDL to overall PON1 activity and function is not clearly established. ApoAI appears of major importance in defining serum PON1 activity and stability, but in the context of an interaction with the phospholipid fraction of HDL. This may involve a role in establishing the architecture of the HDL particle that optimally integrates the PON1 peptide. As the second, major structural peptide of HDL, apoAII may accomplish a similar role. These apolipoproteins, together with others associated with HDL, may also exert a more indirect influence on PON1 function by sequestering oxidised lipids that could compromise enzyme activity. The latter has been exploited therapeutically to give rise to apolipoprotein mimetic peptides that may be useful in limiting oxidative stress within the lipoprotein system, thus permitting PON1 activity to be maximally expressed.

Keywords Apolipoproteins · HDL · ApoAI · ApoAII · ApoJ · LpAI · LpAI · AII

1 Introduction

The role of lipids, and in particular phospholipids, in facilitating paraoxonase-1 (PON1) secretion and its association with high density lipoproteins (HDL) is well established. Indeed, a simple consideration of the structure of PON1 tends to focus attention on the role of lipids. The enzyme retains its highly hydrophobic signal sequence (Sorenson et al., 1999) that consequently requires a hydrophobic environment to enable PON1 to circulate in serum. More importantly, the signal peptide provides the anchor that allows PON1 to associate with HDL, as elegantly

R.W. James (✉)
Department of Internal Medicine, Faculty of Medicine, University of Geneva, Geneva, Switzerland
e-mail: richard.james@hcuge.ch

S.T. Reddy (ed.), *Paraoxonases in Inflammation, Infection, and Toxicology*, Advances in Experimental Medicine and Biology 660, DOI 10.1007/978-1-60761-350-3_16, © Humana Press, a part of Springer Science+Business Media, LLC 2010

demonstrated several years ago (Sorenson et al., 1999). However, HDL is the repository for a number of peptides (apolipoproteins (apo) and enzymes), which fulfil a multitude of functions. As a consequence, they can define the functionality of discrete HDL particles. This review looks at the possible impact of HDL-associated peptides on serum PON1 activity and function, and examines whether any impact is directly or indirectly contributory to its functional efficiency.

2 Apolipoprotein AI

The relationship between PON1 and apoAI, the principal structural peptide of HDL, has been the focus of several studies. Early reports of attempts to purify the enzyme from serum noted the difficulties in removing apoAI from the final product, implying a strong attraction between the two peptides. A landmark study by Sorenson et al. (1999) added to the importance of apoAI for PON1 activity by showing that it was able to stabilise and even increase specific activity when the enzyme was incorporated into phospholipid vesicles. The authors proposed that apoAI and PON1 interacted via the phospholipids present in HDL, rather than direct association of the two peptides. Subsequent studies by Oda et al. (2001) confirmed the ability of apoAI to increase PON1 specific activity. Conversely, they proposed a direct peptide–peptide interaction between apoAI and PON1. The latter was based in part on results from single amino acid mutations in the N-terminal region of apoAI that increased or decreased its ability to improve enzyme specific activity. A feature of their model was a dynamic phase whereby lipidation of apoAI on the HDL particle exposed regions of the apolipoprotein that could interact with and increase PON1 activity. Thus it placed particular importance on the secretion and initial association of the enzyme with HDL as being the phase where apoAI can act to increase PON1 specific activity. With respect to this dynamic model for the impact of apoAI on enzyme activity, a number of studies have since shown that adding purified PON1, or the recombinant enzyme, to synthetic HDL particles (rHDL), or mature HDL isolated from serum, also stabilises activity and improves specific activity. In contemporary studies from our group (Deakin et al., 2002), the primary determinant of PON1 secretion was shown to be the phospholipid component of HDL. ApoAI alone added as an external acceptor had a minor impact on PON1 secretion. Whilst protein-free phospholipid vesicles were able to promote strongly PON1 secretion, the enzyme was not stable and rapidly lost activity. Thus, the optimal complex for stimulating PON1 secretion and maintaining enzyme activity is an apoAI-phospholipid vesicle, or HDL isolated from serum.

There remains the question of a potential role of apoAI in PON1 secretion distinct from its ability to act as an external acceptor for the enzyme. Compelling arguments are the strong affinity between the two peptides and the fact that both are produced and secreted by the liver. One possibility is that apoAI could act as a chaperone, as speculated by Cabana et al. (2003). Two sets of data oppose such a possibility. First, we co-transfected apoAI into our Chinese hamster ovary (CHO) model system of PON1 secretion and showed that co-secretion of apoAI did not influence release of PON1 (Deakin et al., 2002). Secondly, a number of apoAI knockout models show presence of PON1 in serum, despite the absence of apoAI. Complementing these data are observations in patients with apoAI deficiencies, where despite virtual absence of apoAI, there is substantial PON1 activity in serum, albeit significantly reduced (James et al., 1998).

Precisely how apoAI impacts on PON1 enzyme activity has not been fully clarified. One aspect may be modulated distribution of the enzyme within lipoprotein particles. A greater proportion of PON1 was located in the lipid-poor fraction of serum in apoAI knockout mice (Cabana et al., 2003). An alternative or complementary explanation has been provided by studies of Gaidukov and Tawfik (2005) who analysed the affinity of PON1 for synthetic HDL containing different apolipoproteins. The presence of apoAI occasioned an extremely high affinity of PON1 for rHDL, far higher than that provoked by other apolipoproteins. Moreover, the apolipoprotein-mediated affinity of PON1 for HDL correlated positively with enzyme activity and stability. The study also provided evidence for two forms of PON1, stable and unstable. The influence of apolipoproteins on PON1 stability/activity would be mediated by their capacity to transfer the enzyme between the two forms, apoAI being particularly efficient in favouring the stable form. Whether the unstable form of the enzyme represents PON1 in a lipid-poor environment, or bound to forms of HDL (or VLDL (Deakin et al., 2005)) less suitable to PON1 activity, remains to be investigated. In recent studies, Gaidukov et al. (2006) have proposed that PON1 stability may also be influenced by sequence variations of the enzyme, notably the Q192R polymorphism. The latter occurs in a region of the enzyme that, in addition to the retained signal sequence, also mediates association of PON1 with HDL. The Q isoform was found to have a greatly reduced affinity for HDL, giving rise to a less stable form of PON1. The precise role of apoAI remains to be established.

Finally, in a series of elegant studies, Navab and colleagues have described the beneficial effects of apoAI mimetic peptides on serum PON1 (Navab et al., 2007). Such peptides are structurally based on the lipid binding regions of apoAI and have been shown to bind avidly lipid hydroperoxides. The latter tend to accumulate in HDL, where they can inhibit associated enzymes. PON1 is particularly susceptible to inhibition by oxidised lipids. In serum, the mimetic peptides appear to sequester the oxidised lipids, preventing their inhibitory influence on PON1 and leading to an increase in its activity. Can such a function be extrapolated to apoAI? Whilst apoAI may be considerably less efficient than mimetic peptides in sequestering lipid hydroperoxides (Navab et al., 2000), nevertheless it remains an additional avenue by which the apolipoprotein can impact on PON1 activity and stability.

3 Apolipoprotein AII

Despite being quantitatively the second major peptide associated with HDL, the precise function of apoAII remains uncertain (Tailleux et al., 2002). With respect to the two major apolipoproteins, all HDL particles contain apoAI, but a variable percentage contains apoAII in addition to apoAI (Cheung and Albers, 1984). HDL containing only apoAI as the major apolipoprotein are referred to as LpAI, whilst those also containing apoAII are termed LpAI,AII. Thus whilst apoAII has been attributed a structural function, it is not an absolute requirement for formation of HDL. Similar uncertainty surrounds the relationship of apoAII to risk of vascular disease. Initial conclusions were that the apolipoprotein was neutral, or could even increase risk of disease (Tailleux et al., 2002). Recent studies have suggested a more ambivalent association with cardiovascular risk, where the apolipoprotein may bring a measure of protection against the disease (Birjmohun et al., 2007; Winkler et al., 2008). A similar degree of ambiguity characterises the relationship between apoAII and PON1. Early studies showed a positive correlation between the two peptides in serum that was at least as strong as that between apoAI and PON1 (La Du, 1992). An explanation, not totally convincing, was that it reflected the association of both apoAII and PON1 with apoAI. The demonstration that apoAII was associated with only a subfraction of HDL, and the availability of techniques to fractionate and quantitate LpAI and LpAI,AII, allowed us to examine in greater detail the apoAII–PON1 relationship. We observed a significant, positive correlation between apoAII-containing HDL and PON1 arylesterase activity, but no correlation of the enzyme with HDL particles containing apoAI alone (Fig. 1). Further analyses affirmed this dichotomy, as multivariate models showed LpAI,AII to be an independent determinant of serum PON1, whereas LpAI did not enter the multivariate model (Moren et al., 2008). The results contrasted with our previous conclusions, albeit based on quite limited, preliminary studies, that PON1 was essentially associated with HDL containing apoAI but no apoAII (Blatter et al., 1993). Confirmation of the results was provided by showing association of PON1 activity (both arylesterase and paraoxonase) and PON1 peptide (immunologically) with both LpAI and LpAI,AII types of HDL particle (Moren et al., 2008). The major part of serum PON1 was associated with apoAII-containing HDL. Moreover, subsequent analyses suggested a small but significant advantage to the enzyme when associated with LpAI,AII compared to LpAI. Such HDL particles showed a significantly greater capacity to increase and stabilise PON1 secreted from cells, whilst apoAII-associated PON1 was more resistant to inactivation by oxidation. It should

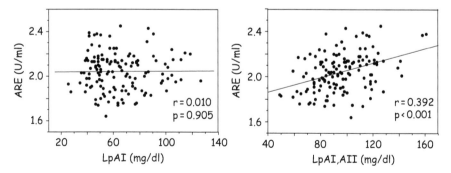

Fig. 1 Correlations between serum concentrations of HDL particles defined by their apoAI and apoAII content and serum PON1 activity

be underlined that in all these studies, apoAII was present in HDL particles also containing apoAI.

How do these results compare with other data in the literature concerning the impact of apoAII on PON1? Firstly, there are no studies that have analysed the association of PON1 with serum HDL particles defined by their apoAI and apoAII complements. Data and conclusions thus refer to in vitro studies or manipulated physiological settings, as in human apoAII transgenic mice. The latter gave rise to HDL with a diminished capacity to protect LDL from oxidation, which correlated with a reduction in the PON1 content of HDL (Cascorbi et al., 1999; Ribas et al., 2004). This is an unsurprising observation, as apoAII is known to displace other apolipoproteins from HDL, and can even displace apoAI. Thus in a pathophysiological situation of excess apoAII, the latter may be generally detrimental to HDL function. However, one cannot conclude from such studies that under physiological conditions, apoAII will also be detrimental to PON1 function. In vitro studies have examined PON1 stability and activity as a function of the peptide composition of HDL. In this context, sophisticated analyses by Gaidukov and Tawfik (2005) are instructive as they clearly establish that synthetic HDL containing apoAI are far superior to synthetic HDL particles containing other apolipoproteins with respect to affinity for PON1 and stimulation of its enzyme activity. Indeed, synthetic HDL containing apoAII alone had no real affinity for PON1 or stimulatory effect on its activity. Whilst these studies clearly underline the importance of apoAI to PON1 activity and function, they do not eliminate possible complementary effects of apoAII on PON1 when present with other apolipoproteins. No studies were performed on synthetic HDL containing both apoAI and apoAII.

How could the physiological impact of apoAII on HDL benefit PON1? It is well established that addition of apoAII to human HDL influences the metabolism of the lipoprotein, where it is also known to stabilise HDL structure (Boucher et al., 2004). PON1 shows an extreme affinity for HDL-type lipid complexes that furnish a physico-chemical configuration, as yet undefined, well suited to enzyme stability and activity. Thus one could speculate that apoAII favours and stabilises a particular type of HDL particle that optimises that stimulatory effect of apoAI on PON1 function.

Finally, we have examined whether the apoAII association of PON1 could have clinical relevance. To this end, we analysed the PON1 distribution between LpAI and LpAI,AII in type 2 diabetic patients. Diabetes is known to influence HDL composition and function, which appears to be a contributory factor to the greatly increased risk of coronary disease in this population. Preliminary data from 12 patients and matched non-diabetic controls are shown in Table 1. The groups were well matched for age, sex and most lipids. ARE activity was higher in the nondiabetic group, as has been observed in other studies, albeit non-significant probably due to the small number of participants. There was, however, a marked difference in the distribution of PON1 activity between the two types of HDL particle. Diabetic patients had a higher percentage of activity associated with the LpAI subfraction. As concentrations of apoAI and apoAII were comparable between the two groups, they cannot explain the observation. The latter is nevertheless intriguing given that our

Table 1 Distribution of PON1 arylesterase activity between HDL particles containing apoAI alone (LpAI) or apoAI + apoAII (LpAI,AII)

Parameter	Control	Type 2 diabetes	p
n (M/F)	12 (8/4)	12 (7/5)	ns
Age (y)	56.7 ± 10.3	57.2 ± 11.4	ns
Cholesterol (mm/L)	5.36 ± 0.74	4.44 ± 0.84	0.014
Triglycerides (mm/L)	1.03 ± 0.31	1.41 ± 0.74	0.10
HDL-cholesterol (mm/L)	1.22 ± 0.24	1.16 ± 0.19	ns
ApoAI (g/L)	1.61 ± 0.35	1.57 ± 0.33	ns
ApoAII (g/L)	0.47 ± 0.09	0.46 ± 0.22	ns
Serum ARE (OD/μl)	7.43 ± 1.14	6.78 ± 1.49	ns
% ARE activity in LpAI	44.2 ± 8.0	56.9 ± 9.5	0.003

ARE, arylesterase activity; OD, optical density

previous studies (Moren et al., 2008) suggested that PON1 associated with LpAI was less well protected from stress. Although preliminary, the data indicate that the distribution of PON1 between different types of HDL particle appears to be of clinical relevance.

4 Apolipoprotein J

ApoJ (or clusterin) is a ubiquitously synthesised peptide associated with HDL in serum. Its precise function is unclear, but it has a chaperone-type activity and is implicated in regulation of cell death. Oxidative stress appears to play a prominent role in regulating its expression (Trougakos and Gonos, 2006). In early studies we demonstrated that apoJ was in part associated to HDL particles also containing PON1 (Blatter et al., 1993). The observation was confirmed and extended by Kelso et al. who showed a high affinity of apoJ for purified PON1 (Kelso et al., 1994). Given the extremely hydrophobic nature of apoJ, its binding of purified PON1, with its hydrophobic signal peptide, is not unexpected. Reciprocal changes in serum apoJ and PON1 were observed in atherosclerosis-susceptible, transgenic murine models subjected to an atherogenic diet or treated with oxidatively modified LDL: significant increases in apoJ were accompanied by significant decreases in PON1 (Navab et al., 1997). Other studies have shown the presence of PON1 and apoJ, together with apoAI, during the development and progression of aortic plaques in man (Mackness et al., 1997). Whether they may have a common origin from an HDL complex containing all three peptides was unclear. Indeed, it is unlikely, at least for apoJ given the widespread tissue expression of this peptide, especially under conditions of increased cell death.

None of these studies provide any evidence for a functional interaction between PON1 and apoJ under physiological conditions. However, apoJ is a highly hydrophobic peptide with strong lipid-binding properties. Thus there is the potential to bind and sequester oxidised lipids and shield PON1 from the detrimental effects of oxidative stress. Such a functional interaction between apoJ and PON1 has not been investigated (see section on mimetic peptides).

5 Mimetic Peptides

The ability to sequester oxidised lipids was shown some two decades ago to be a powerful means of reducing the atherogenic potential of LDL, and limiting the ability of cells to render LDL atherogenic (Navab et al., 2004). ApoAI was initially shown to be effective in this context, where the lipid-binding amphipathic helix integral to apolipoprotein structure plays a dominant role. Realisation of the lipid-binding importance of the α-helical structures led to the investigation of mimetic peptides containing such structures (Segrest et al., 1992). The latter are short sequences, up to 18 amino acids, that form the class A amphipathic helix characteristic of apolipoproteins, without necessarily having sequence homology with apolipoproteins. With respect to an impact on PON1, data exist on mimetic peptides based on three apolipoproteins: apoAI (Navab et al., 2004), apoJ (Navab et al., 2005b) and apoE (Gupta et al., 2005). ApoAI mimetic peptides have been the most widely analysed, with a variety of structures differing in their affinity for lipids. The mimetics have been shown to reduce the extent of lesion formation in animal models of atherosclerosis, notably with a general beneficial impact on HDL function (Navab et al., 2004). The latter included an improved anti-oxidant/anti-inflammatory capacity, with an increase in the activity of serum PON1 (Navab et al., 2004; 2005a). Comparable results were obtained with apoJ mimetics, and even short peptide sequences unable to form an α-helix were found to be effective (Navab et al., 2005b) in favourably modulating HDL function, increasing PON1 activity and limiting lesion development. The studies have also been extended to apoE mimetic peptides, with equivalent results regarding improved HDL function and significantly increased PON1 activity (Gupta et al., 2005).

In all cases the efficacy of the mimetics appears to lie in their high affinity for oxidised lipids. By effectively scavenging such lipids, the mimetics prevent inhibition of enzyme activity, not only of PON1, which is highly susceptible to oxidative stress, but also of other HDL-associated enzymes (Forte et al., 2002). The particularly high affinity for oxidised lipids effectively neutralises their inhibitory actions, such that PON1 remains active even in the presence of the mimetic peptides. This contrasts with the native apolipoproteins that can also bind lipids via their α-helical structures, but less effectively shield PON1 from the inhibitory effects of the complexed, oxidised lipids. Nevertheless, given that apolipoproteins tend to be present in far greater concentrations than PON1, either collectively, or individually (apoAI), there is the potential for such apolipoproteins to impact on PON1 activity by acting as scavengers of oxidised lipids.

6 Conclusions

Available data indicate that apolipoproteins make a significant contribution to serum PON1 activity and functional efficiency. It principally concerns apoAI, which increases the affinity of vesicles for PON1 and improves enzyme activity and stability. As there is no proof of a direct interaction between PON1 and apoAI, the contribution of apoAI may lie in its capacity to produce a vesicle (HDL) architecture optimal for PON1 activity. In this context, one could also envisage a role for apoAII possibly via interaction with apoAI, as it also impacts on HDL structure. Finally, a more indirect contribution of these plus other apolipoproteins to PON1 function can be proposed based on the pioneering work with apolipoprotein mimetic peptides and their capacity to neutralise the inhibitory effects of oxidised lipids on PON1 activity.

References

Birjmohun RS, Dallinga-Thie GM, Kuivenhoven JA et al. (2007). Apolipoprotein A-II is inversely associated with risk of future coronary artery disease. Circulation 116: 2029–2035.

Blatter M-C, James RW, Messmer S et al. (1993). Identification of a distinct human high-density lipoprotein subspecies defined by a lipoprotein-associated protein, K-45. Identity of K-45 with paraoxonase. Eur J Biochem 211:871–879.

Boucher J, Ramsamy TA, Braschi S et al. (2004). Apolipoprotein A-II regulates HDL stability and affects hepatic lipase association and activity. J Lipid Res 45:849–858.

Cabana VG, Reardon CA, Feng N et al. (2003). Serum paraoxonase: effect of the apolipoprotein composition of HDL and the acute phase response. J Lipid Res 44:780–792.

Cascorbi I, Laule M, Mrozikiewicz PM et al. (1999). Mutations in the human paraoxonase 1 gene: frequencies, allelic linkages, and association with coronary artery disease. Pharmacogenet 9:755–761.

Cheung MC, Albers JJ (1984). Characterization of lipoprotein particles isolated by immunoaffinity chromatography: particles containing A-I and A-II and particles containing A-I but no A-II. J Biol Chem 259:12201–12209.

Deakin S, Leviev I, Gomaraschi M et al. (2002). Enzymatically active paraoxonase-1 is located at the external membrane of producing cells and released by a high affinity, saturable, desorption mechanism. J Biol Chem 277:4301–4308.

Deakin S, Moren X, James RW (2005). Very low density lipoproteins provide a vector for secretion of paraoxonase-1 from cells. Atherosclerosis 179:17–25.

Forte TM, Subbanagounder G, Berliner JA et al. (2002). Altered activities of anti-atherogenic enzymes, LCAT, paraoxonase, and platelet activating factor acetylhydrolase in atherosclerosis-susceptible mice. J Lipid Res 43:477–485.

Gaidukov L, Rosenblat M, Aviram M et al. (2006). The 192R/Q polymorphs of serum paraoxonase PON1 differ in HDL binding, lipolactonase stimulation, and cholesterol efflux. J Lipid Res 47:2492–2502.

Gaidukov L, Tawfik DS (2005). High affinity, stability, and lactonase activity of serum paraoxonase PON1 anchored on HDL with ApoA-I. Biochemistry 44:11843–11854.

Gupta H, White CR, Handattu S et al. (2005). Apolipoprotein E mimetic peptide dramatically lowers plasma cholesterol and restores endothelial function in watanabe heritable hyperlipidemic rabbits. Circulation 111:3112–3118.

James RW, Blatter Garin MC, Calabresi L et al. (1998). Modulated serum activities and concentrations of paraoxonase in high density lipoprotein deficiency states. Atherosclerosis 139:77–82.

Kelso GJ, Stuart WD, Richter RJ et al. (1994). Apolipoprotein J is associated with paraoxonase in human plasma. Biochemistry 33:832–839.

La Du BN (1992). Human serum paraoxonase/arylesterase. Pharmacogenetics of drug metabolism. W. Kalow (ed.). New York: Pergamon Press 51–91.

Mackness B, Hunt R, Durrington PN et al. (1997). Increased immunolocalization of paraoxonase, clusterin, and apolipoprotein A-I in the human artery wall with the progression of atherosclerosis. Arterioscler Thromb Vasc Biol 17:1233–1238.

Moren X, Deakin S, Liu M-L et al. (2008). HDL subfraction distribution of paraoxonase-1 and its relevance to enzyme activity and resistance to oxidative stress. J Lipid Res 49:1228–1234.

Navab M, Anantharamaiah GM, Reddy ST et al. (2005a). Oral small peptides render HDL anti-inflammatory in mice and monkeys and reduce atherosclerosis in apoE null mice. Circulation Res 97:524–532.

Navab M, Anantharamaiah GM, Reddy ST et al. (2007). Peptide mimetics of apolipoproteins improve HDL function. J Clin Lipidol 1:142–147.

Navab M, Anantharamaiah GM, Reddy ST et al. (2004). Human apolipoprotein A-I and A-I mimetic peptides: potential for atherosclerosis reversal. Curr Opin Lipidol 15:645–649.

Navab M, Anantharamaiah GM, Reddy ST et al. (2005b). An oral apoJ peptide renders HDL anti-inflammatory in mice and monkeys and dramatically reduces atherosclerosis in apolipoprotein E-null mice. Arterioscler Thromb Vasc Biol 25:1932–1937.

Navab M, Hama-Levy S, Van Lenten B et al. (1997). Mildly oxidised LDL induces an increased apolipoprotein J/paraoxonase ratio. J Clin Invest 99:2005–2019.

Navab M, Hama SY, Cooke CJ et al. (2000). Normal high density lipoprotein inhibits three steps in the formation of mildly oxidised low density lipoprotein: step 1. J Lipid Res 41:1481–1494.

Oda MN, Bielicki JK, Berger T et al. (2001). Cysteine substitutions in apolipoprotein A-I primary structure modulate paraoxonase activity. Biochemistry 40:1710–1718.

Ribas V, Sanchez-Quesada JL, Anton R et al. (2004). Human apolipoprotein A-II enrichment displaces paraoxonase from HDL and impairs its antioxidant properties: a new mechanism linking HDL protein composition and antiatherogenic potential. Circ Res 95:789–797.

Segrest JP, Jones MK, De Loof H et al. (1992). The amphipathic helix in the exchangeable apolipoproteins: a review of secondary structure and function. J Lipid Res. 33:141–166.

Sorenson RC, Bisgaier CL, Aviram M et al. (1999). Human serum paraoxonase/arylesterase's retained hydrophobic N-terminal leader sequence associates with HDLs by binding phospholipids: apolipoprotein A-I stabilizes activity. Arterioscler Thromb Vasc Biol 19:2214–2225.

Tailleux A, Duriez P, Fruchart JC et al. (2002). Apolipoprotein A-II, HDL metabolism and atherosclerosis. Atherosclerosis 164:1–13.

Trougakos IP, Gonos ES (2006). Regulation of clusterin/apolipoprotein J, a functional homologue to the small heat shock proteins, by oxidative stress in ageing and age-related diseases. Free Radic Res 40:1324–1334.

Winkler K, Hoffmann MM, Seelhorst U et al. (2008). Apolipoprotein A-II is a negative risk indicator for cardiovascular and total mortality: findings from the Ludwigshafen Risk and Cardiovascular Health Study. Clin Chem 54:1405–1406.

Paraoxonase 1, Quorum Sensing, and *P. aeruginosa* Infection: A Novel Model

M.L Estin, D.A Stoltz, and J. Zabner

Time flies like an arrow; fruit flies like a banana
–Groucho Marx

Abstract *Pseudomonas aeruginosa* is a Gram-negative bacterium which exacts a heavy burden on immunocompromised patients, but is non-pathogenic in a healthy host. Using small signaling molecules called acyl-homoserine lactones (AHLs), populations of *P. aeruginosa* can coordinate phenotypic changes, including biofilm formation and virulence factor secretion. This concentration-dependent process is called quorum sensing (QS). Interference with QS has been identified as a potential source of new treatments for *P. aeruginosa* infection. The human enzyme paraoxonase 1 (PON1) degrades AHL molecules, and is a promising candidate for QS interference therapy. Although paraoxonase orthologs exist in many species, genetic redundancy in humans and other mammals has made studying the specific effects of PON1 quite difficult. Arthropods, however, do not express any PON homologs. We generated a novel model to study the specific effects of PON1 by transgenically expressing human PON1 in *Drosophila melanogaster*. Using this model, we showed that *P. aeruginosa* infection lethality is QS-dependent, and that expression of PON1 has a protective effect. This work demonstrates the value of a *D. melanogaster* model for investigating the specific functions of members of the paraoxonase family in vivo, and suggests that PON1 plays a role in innate immunity.

Keywords Paraoxonase · *Pseudomonas aeruginosa* · *Drosophila melanogaster* · Quorum sensing · 3OC12-HSL · Biofilms · Airway epithelium · Host defense · Evolution

J. Zabner (✉)
Department of Internal Medicine, University of Iowa, Iowa City, IA 52242, USA
e-mail: joseph-zabner@uiowa.edu

S.T. Reddy (ed.), *Paraoxonases in Inflammation, Infection, and Toxicology*, Advances in Experimental Medicine and Biology 660, DOI 10.1007/978-1-60761-350-3_17, © Humana Press, a part of Springer Science+Business Media, LLC 2010

1 Introduction

Pseudomonas aeruginosa is a Gram-negative bacterium which is ubiquitous in soil and water. Though generally non-pathogenic in a healthy host, *P. aeruginosa* represents a major cause of morbidity and mortality in immunocompromised patients. Burn victims, patients with diabetic foot ulcers, and those with chronic lung diseases such as cystic fibrosis (CF) are especially at risk of infection.

The ability of *P. aeruginosa* to form a biofilm contributes to antibiotic resistance and to poor outcomes in these populations. In planktonic cultures, *P. aeruginosa* is susceptible to a number of treatments. Within a biofilm, however, the bacteria are shielded from host immune responses and are much harder to treat; protected by a layer of secreted peptidogylcans, bacteria can more easily participate in horizontal gene transfer, and may evade drugs that target replication by entering a dormant phase (Fux et al., 2005). In *P. aeruginosa*, biofilm formation and virulence factor expression are regulated by quorum sensing.

Quorum sensing (QS) is a density-dependent process by which a population of bacteria can coordinate phenotypic changes by synthesizing, secreting, and responding to small signaling molecules known as autoinducers, or pheromones. QS was first identified in *Vibrio fischeri,* a marine bacterium which symbiotically colonizes the light organs of certain fish and squid (Waters and Bassler, 2005). Bacterial growth in this niche can become very dense, and when it reaches a threshold density, *V. fischeri* begins to transcribe luciferase and becomes biolumescent (Nealson, 1977). This behavior is accomplished via a positive feedback loop involving the production and detection of autoinducers, or quorum-sensing signals. Both *V. fischeri* and *P. aeruginosa* use acyl-homoserine lactones (AHLs) as their autoinducers.

Depending on the length of the carbon tail, AHLs can diffuse actively or passively across cell membranes (Pearson et al., 1999) in order to activate transcription factors in the LuxR family, which regulate certain target genes (Camilli and Bassler, 2006). AHL signaling is common in Gram-negative bacteria, but Gram-positive bacteria generally use short oligopeptide signals comprised of 5–17 amino acids. These signals do not diffuse freely through membranes, but may bind to cell-surface receptors. Finally, both Gram-negative and Gram-positive bacteria have been shown to produce and respond to a third category of quorum-sensing signal, known as autoinducer two (AI-2). As a result, AI-2 may facilitate "crosstalk" between different bacterial species (Lowery et al., 2008; Vlamakis and Kolter, 2005). Table 1 summarizes the three types of autoinducer molecules.

Quorum sensing signals/autoinducers

Pseudomonas aeruginosa has been shown to respond to both AHL and AI-2 signaling, but cannot synthesize AI-2 (Duan et al., 2003); we will therefore focus on the AHL pathway. AHL signaling in *P. aeruginosa* is controlled by a two-component regulatory system, which is depicted in Fig. 1. The AHL autoinducer *N*-3-oxodocecanoyl homoserine lactone (3OC12-HSL) diffuses across the cell membranes and activates the transcription factor LasR, which triggers a positive feedback loop by initiating the transcription of LasI, an enzyme which produces 3OC12-HSL. LasR also turns on the RhlR/RhlI regulatory circuit. The RhlI gene encodes an enzyme which produces another AHL, *N*-butanoyl homoserine lactone

Table 17.1 Quorum sensing signals/autoinducers

Acyl-homoserine lactone (AHL)	Autoinducer-2 (AI-2)	Oligopeptide autoinducer
• Gram-negative bacteria • Active or passive diffusion through cell membranes • Binds LuxR family transcription factors	• Gram-negative and Gram-positive bacteria • Imported into cytoplasm by specialized proteins or remains outside cell • Binds LuxS family transcription factors or cell-surface receptors	• Gram-positive bacteria • Does not diffuse through cell membranes • Binds cell-surface receptors
(core structure) (R group example: *P. aeruginosa* 3OC12-HSL)	(Example: *V. harveyi*)	(Example: *S. aureus*)

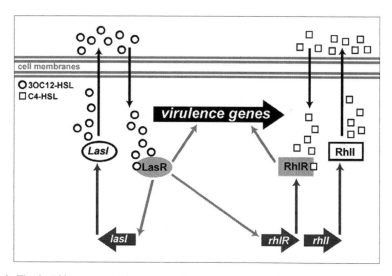

Fig. 1 The *las/rhl* quorum-sensing system. *P. aeruginosa* uses a two-component QS system to regulate virulence factor expression and biofilm formation. The 3OC12-HSL autoinducer diffuses across cell membranes and activates the transcription factor LasR, which initiates the transcription of LasI, an enzyme which produces 3OC12-HSL. LasR also activates the *rhlR* and *rhlI* genes. RhlI is an enzyme that produces C4-HSL, the autoinducer associated with the *rhlI/rhlR* circuit. C4-HSL activates RhlR, a transcription factor encoded by the *rhlR* gene. Upon activation, RhlR initiates transcription of *rhlI*. Both LasR and RhlR also activate genes associated with virulence factors and biofilm formation

(C4-HSL); RhlR is a transcription factor which, when activated by C4-HSL, turns on transcription of RhlI. Numerous other genes, perhaps as much as 6% of the approximately 6,000 encoded by *P. aeruginosa* (Schuster et al., 2003; Wagner et al., 2003; Whiteley et al., 1999), are targeted by LasR and RhlR, either independently or together. Among these genes are those related to biofilm formation, and those that encode numerous virulence factors, including exotoxin A, alkaline protease, superoxide dismutase, elastase, and pyocyanin (Smith and Iglewski, 2003). In vivo experiments in animal models of wound and pulmonary infection have demonstrated attenuated virulence in QS-deficient strains of *P. aeruginosa* (Imamura et al., 2005; Pearson et al., 2000; Rumbaugh et al., 1999; Tang et al., 1996; Wu et al., 2001).

Interference with QS signaling, or "quorum quenching," has been identified in nature and provides a promising avenue for new therapies against a range of bacterial pathogens. For example, soft-rot infection by the AHL quorum-sensing bacterium *Erwinia carotovora* was inhibited when AiiA, a lactonase naturally produced by *Bacillus* sp. bacteria, was transgenically expressed in tobacco and cabbage plants (Dong et al., 2001; Dong et al., 2000). Similarly, treatment with a synthetic analog of the furanones produced by the algae *Delisea pulchra* was shown to improve survival and clearance of pulmonary *P. aeruginosa* infection in a murine model (Hentzer et al., 2003; Wu et al., 2004).

The paraoxonase family of enzymes, including PON1, PON2, and PON3, has been associated with multiple enzymatic activities. In humans and other mammals, PON1 is secreted into the blood. It is capable of degrading paraoxon and other organophosphates, is active as a lactonase, and has been shown to protect against lipid oxidation (Draganov et al., 2005). Although the native substrate(s) of the PON family remain unknown, all three hydrolyze lactones. Recent work suggests that this may be PON's endogenous enzymatic role (Khersonsky and Tawfik, 2005).

PON1 has been proposed as a candidate for *P. aeruginosa* QS signaling interference. Our laboratory has previously shown that human and murine airway epithelial cells, which endogenously express all three PONs, degrade 3OC12-HSL (Chun et al., 2004). In addition, mouse serum-PON1 degrades 3OC12-HSL in vitro (Ozer et al., 2005). Although *P. aeruginosa* QS signaling was not enhanced in mice with a PON1 deficiency, mRNA expression levels suggest that this may be due to compensation by other members of the PON family (Ozer et al., 2005). The redundancy of PON in humans and other mammals has made it difficult to tease apart the specific activities of individual members of the family in existing models. This chapter will describe a novel *D. melanogaster* model for PON1, and will summarize our recent data.

2 Paraoxonases are Widely Conserved, but Have Not Been Found in Arthropods

Previous work has shown that the paraoxonases are highly conserved among mammals, and indeed across many species (Draganov and La Du, 2004; Yang et al., 2005). These studies further suggest that redundancy in the paraoxonase family is

Fig. 2 PON1 phylogeny. Number indicates number of species with a known or predicted PON homolog in the NCBI and/or Superfamily 1.73 databases

Eukaryotes
   Animals
      Chordates *(25)*
         Vertebrates *(24)*
            Mammals *(15)*
            Bony fishes *(4)*
            Amphibians *(2)*
            Birds *(3)*
         Lancelets *(1)*
      Nematodes *(4)*
      Echinoderms *(1)*
      Cnidarians *(1)*
      Arthropods *(0)*
   Fungi
      Ascomycetes *(20)*
      Basidomycetes *(3)*
Bacteria
   Proteobacteria
      Alphaproteobacteria (14)
      Betaproteobacteria (4)
      Gammaproteobacteria (11)
   Actinobacteria *(11)*
   CFB group bacteria *(5)*
   Planctomycetes *(3)*
   Spirochetes *(2)*
   Cyanobacteria *(1)*
   GNS bacteria *(1)*
   Lentisphaerales *(1)*
   Verrucomicrobia *(1)*
   Undefined *(1)*
Archaea *(2)*

the result of gene duplication, and that PON2 is the oldest member of the family while PON1 is the youngest. Interestingly, PON2 is primarily active as a lactonase, supporting the notion that lactonase activity is among the oldest endogenous functions of PON species (Draganov and La Du, 2004).

We generated a systematic phylogeny of PON orthologs based on identified and predicted amino acid reference sequences in the NCBI/BLASTp (Altschul et al., 1997) and Superfamily 1.73 (Gough et al., 2001) databases. Figure 2 shows the number of species per Domain, Phylum, Class, or Order which generated a match in these databases. PON-like domains are conserved across all three major Domains: Archaea, Bacteria, and Protozoa. We next generated an amino acid alignment using the T-Coffee program, and examined several important functional sites within the PON1 sequence (Labarga et al., 2007). We found that these sites are generally conserved as well. Figure 3 identifies key glycosylation sites and active site residues thought to be important for lactonase activity. The PON family is catalytically promiscuous, and has evolved to develop divergent enzymatic functions in many species (Khersonsky and Tawfik, 2006). These data suggest that PON possesses

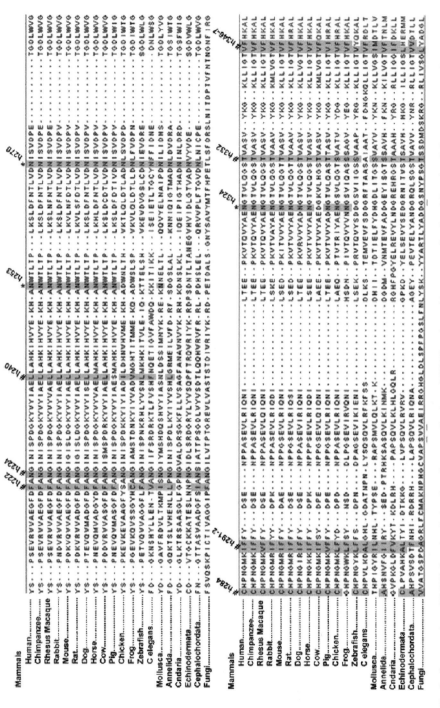

Fig. 3 Conservation of PON1 functional sites in eukaryotes. Portion of PON1 amino acid sequence. Predicted functional sites are conserved across many species. Highlighted sites are conserved, according to the ClustalX algorithm. [*] indicates glycosylation site, and [#] indicates key active site residue; numbers are given according to the human PON1 amino acid sequence

a long evolutionary history. Finally, although PON orthologs are present in most animals, it is not able that PON orthologs are absent from all arthropod sequences in both databases, including that of *Drosophila melanogaster*.

3 Generating a PON1 Transgenic *Drosophila melanogaster* Infection Model

The finding that insects express no PON orthologs suggested that *D. melanogaster* might provide an ideal model system to study the distinct functions of individual members of the PON family. The *D. melanogaster* model has other experimental advantages as well, including a short generation time and well-characterized genetics. The *D. melanogaster* genome contains orthologs for as much as 80% of human disease genes (Gilbert, 2008), and it has been an important model for the discovery of key components of human innate immunity, including the Toll, Imd, and JAK/STAT pathways (Lemaitre and Hoffmann, 2007).

We assessed *D. melanogaster* as a model for *P. aeruginosa* infection. Using a previously-described abdomen wound infection model (Avet-Rochex et al., 2005; D'Argenio et al., 2001; Lau et al., 2003), we first showed that *P. aeruginosa* infection lethality in wild-type *D. melanogaster* is QS-dependent. Briefly, we dipped a needle in a suspension of *P. aeruginosa* expressing green fluorescent protein (GFP) under the control of a QS-dependent promoter, and pricked the fly's abdomen. Intense GFP expression 18 h post-inoculation indicated that *P. aeruginosa* was indeed expressing QS-dependent genes.

Next, we showed that QS increases *P. aeruginosa* lethality by infecting with mutant *P. aeruginosa* strains deficient in either AHL production (Δ*lasI/rhlI*) or detection (Δ*lasR/rhlR*). Compared to controls inoculated with wild-type *P. aeruginosa*, flies inoculated with either mutant strain had greater rates of survival. Interestingly, feeding 3OC12-HSL and C4-HSL to flies restored *P. aeruginosa* virulence in those infected with the AHL production mutant (Δ*lasI/rhlI*) but not the detection mutant (Δ*lasR/rhlR*).

Finally, we generated a novel PON whole-organism model in order to test whether the lactonase activity of PON1 would confer a protective effect against *P. aeruginosa* infection in vivo. We generated transgenic *D. melanogaster* ubiquitously expressing human PON1 (hPON1) using the GAL4-UAS system with a *da* promoter (Brand and Perrimon, 1993).

4 PON1 Transgenic Flies are Protected from Organophosphate Poisoning

After confirming PON1 expression and protein production by qPCR and Western blot, we set out to validate the in vivo function of this protein. The ability of PON1 to degrade organophosphates was among the first enzymatic activities identified for

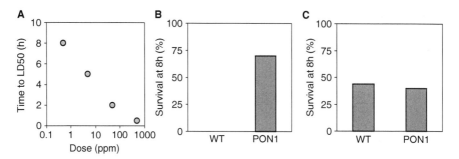

Fig. 4 Organophosphate lethality. Transgenic expression of human PON1 protects *D. melanogaster* from organophosphate toxicity. (**a**) Chlorpyrifos toxicity is dose-dependent in wild-type flies. (**b**) PON1 flies are protected from chlorpyrifos toxicity. (**c**) Similar DSM lethality is observed in PON1 and wild-type flies

this family. We were therefore able to measure PON1 activity by exposing both transgenic and wild-type flies to the organophosphate chlorpyrifos, and comparing rates of survival between the two groups. Chlorpyrifos lethality was dose-dependent (Fig. 4a), and survival rates were significantly higher among PON1 transgenic (UAS-PON1/*da*-GAL4) flies than among (da-GAL4 +/+) controls (Fig. 4b). We also conducted control experiments with an organophosphate not degraded by PON1, demeton-S-methyl (DSM). As shown in Fig. 4c, DSM survival was similar in both groups.

5 PON1 Expression Protects Against *Pseudomonas aeruginosa* Lethality

Using the abdominal wound infection model we previously validated, we demonstrated that PON1 improved survival after infection with the PA01 strain of *P. aeruginosa*. We showed similar results with two clinical isolates of *P. aeruginosa*. Survival rates were similar to those of control flies infected with the AHL production-mutant Δ*lasI/rhlI*. Moreover, the PON1 transgenic flies were also protected against infection with *Serratia marcescens,* another AHL-sensing bacterium which is lethal to flies. However, when flies were infected with *Staphylococcus aureus*, which does not use AHL-mediated QS, the PON1 transgenic and control populations showed identical survival rates. Figure 5 summarizes this data (Harel et al., 2004; Tavori et al., 2008).

6 Concluding Remarks

The paraoxonases are a versatile family of enzymes, which have developed divergent catalytic specificities in many species over a long evolutionary history. Recent evidence suggests that the endogenous role of paraoxonase may be as a lactonase,

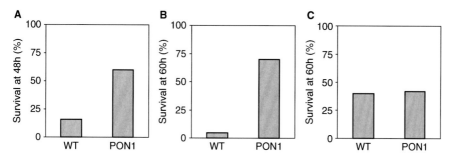

Fig. 5 Bacterial infection lethality. Transgenic expression of human PON1 protects *D. melanogaster* from the Gram-negative, AHL-sensing bacteria (**a**) *P. aeruginosa* and (**b**) *S. marcescens*, but not from (**c**) *S. aureus*, which is Gram-positive and does not use AHLs for quorum-sensing

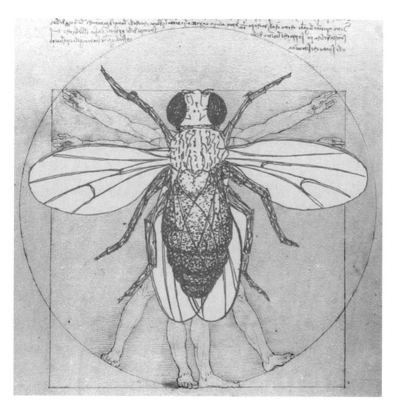

Fig. 6 Transgenic expression of a human enzyme in *D. melanogaster*. This novel whole-organism model allowed us to investigate the specific effects of a single member of the human paraoxonase family on innate immunity

and that this is the primary activity of PON2, the oldest member of the PON family. PON1 is the youngest member of the family. It is present in many tissues, and is secreted into the blood in humans and other mammals (Marsillach et al., 2008). PON1 is also active as a lactonase, but also degrades paraoxons. In vitro evidence has suggested that the former function may allow PON1 to play a role in host defense against infection by AHL-sensing bacteria. However, confirmation of these PON1-specific observations in vivo has been difficult due to genetic redundancy in humans and other mammals.

Taking advantage of the unique absence of paraoxonase orthologs in *D. melanogaster* and other arthropods, we developed a novel whole-organism model in which to investigate the specific effects of human PON1 on *P. aeruginosa* infection and QS. We expressed human PON1 in *D. melanogaster* and confirmed PON1 mRNA and protein expression in our transgenic model with biochemical assays, and demonstrated in vivo activity using chlorpyrifos susceptibility as a functional end point.

Using an abdominal wound infection model, we showed that *P. aeruginosa* lethality is dependent on QS, and that transgenic expression of hPON1 significantly increases survival after inoculation. PON1 was similarly protective after infection with another AHL-sensing bacterium, *S. marcescens*, but did not improve the survival rates of flies inoculated with *S. aureus,* a Gram-positive bacterium which does not use AHLs for QS. Taken together, our data suggest that PON1 may play an important role in innate immunity, particularly in preventing Gram-negative bacterial infection and biofilm formation. Moreover, we showed that *D. melanogaster* is a valuable model for investigating the specific functions of members of the paraoxonase family in vivo (see Fig. 6).

References

Altschul, S. F., Madden, T. L., Schaffer, A. A., Zhang, J., Zhang, Z., Miller, W., Lipman, D. J. (1997) *Nucleic Acids Res* **25**, 3389–3402.
Avet-Rochex, A., Bergeret, E., Attree, I., Meister, M., Fauvarque, M. O. (2005) *Cell Microbiol* **7**, 799–810.
Brand, A. H., Perrimon, N. (1993) *Development* **118**, 401–415.
Camilli, A., Bassler, B. L. (2006) *Science* **311**, 1113–1116.
Chun, C. K., Ozer, E. A., Welsh, M. J., Zabner, J., Greenberg, E. P. (2004) *Proc Natl Acad Sci USA* **101**, 3587–3590.
Dong, Y. H., Wang, L. H., Xu, J. L., Zhang, H. B., Zhang, X. F., Zhang, L. H. (2001) *Nature* **411**, 813–817.
Dong, Y. H., Xu, J. L., Li, X. Z., Zhang, L. H. (2000) *Proc Natl Acad Sci USA* **97**, 3526–3531.
Draganov, D. I., La Du, B. N. (2004) *Naunyn Schmiedebergs Arch Pharmacol* **369**, 78–88.
Draganov, D. I., Teiber, J. F., Speelman, A., Osawa, Y., Sunahara, R., La Du, B. N. (2005) *J Lipid Res* **46**, 1239–1247.
Duan, K., Dammel, C., Stein, J., Rabin, H., Surette, M. G. (2003) *Mol Microbiol* **50**, 1477–1491.
D'Argenio, D. A., Gallagher, L. A. , Berg, C. A., Manoil, C. (2001) *J Bacteriol* **183**, 1466–1471.
Fux, C. A., Costerton, J. W., Stewart, P. S., Stoodley, P. (2005) *Trends Microbiol* **13**, 34–40.
Gilbert, L. I. (2008) *Mol Cell Endocrinol* **293**, 25–31.
Gough, J., Karplus, K., Hughey, R., Chothia, C. (2001) *J Mol Biol* **313**, 903–919.
Harel, M., Aharoni, A., Gaidukov, L., Brumshtein, B., Khersonsky, O., Meged, R., Dvir, H., Ravelli, R. B., McCarthy, A., Toker, L., et al. (2004) *Nat Struct Mol Biol* **11**, 412–419.
Hentzer, M., Wu, H., Andersen, J. B., Riedel, K., Rasmussen, T. B., Bagge, N., Kumar, N., Schembri, M. A., Song, Z., Kristoffersen, P., et al. (2003) *Embo J* **22**, 3803–3815.

Imamura, Y., Yanagihara, K., Tomono, K., Ohno, H., Higashiyama, Y., Miyazaki, Y., Hirakata, Y., Mizuta, Y., Kadota, J., Tsukamoto, K., et al. (2005) *J. Med. Microbiol* **54**, 515–518.

Khersonsky, O., Tawfik, D. S. (2005) *Biochemistry* **44**, 6371–6382.

Khersonsky, O., Tawfik, D. S. (2006) *Chembiochem* **7**, 49–53.

Labarga, A., Valentin, F., Anderson, M., Lopez, R. (2007) *Nucleic Acids Res* **35**, W6–11.

Lau, G. W., Goumnerov, B. C., Walendziewicz, C. L., Hewitson, J., Xiao, W., Mahajan-Miklos, S., Tompkins, R. G., Perkins, L. A., Rahme, L. G. (2003) *Infect Immun* **71**, 4059–4066.

Lemaitre, B., Hoffmann, J. (2007) *Annu Rev Immunol* **25**, 697–743.

Lowery, C. A., Dickerson, T. J., Janda, K. D. (2008) *Chem Soc Rev* **37**, 1337–1346.

Marsillach, J., Mackness, B., Mackness, M., Riu, F., Beltran, R., Joven, J., Camps, J. (2008) *Free Radic Biol Med* **45**, 146–157.

Nealson, K. H. (1977) *Arch Microbiol* **112**, 73–79.

Ozer, E. A., Pezzulo, A., Shih, D. M., Chun, C., Furlong, C., Lusis, A. J., Greenberg, E. P., Zabner, J. (2005) *FEMS Microbiol Lett* **253**, 29–37.

Pearson, J. P., Feldman, M., Iglewski, B. H., Prince, A. (2000) *Infect Immun* **68**, 4331–4334.

Pearson, J. P., Van Delden, C., Iglewski, B. H. (1999) *J Bacteriol* **181**, 1203–1210.

Rumbaugh, K. P., Griswold, J. A., Iglewski, B. H., Hamood, A. N. (1999) *Infect. Immun* **67**, 5854–5862.

Schuster, M., Lostroh, C. P., Ogi, T., Greenberg, E. P. (2003) *J. Bacteriol* **185**, 2066–2079.

Smith, R. S., Iglewski, B. H. (2003) *Curr Opin Microbiol* **6**, 56–60.

Tang, H. B., DiMango, E., Bryan, R., Gambello, M., Iglewski, B. H., Goldberg, J. B., Prince, A. (1996) *Infect Immun* **64**, 37–43.

Tavori, H., Khatib, S., Aviram, M., Vaya, J. (2008) *Bioorg Med Chem* **16**, 7504–7509.

Vlamakis, H. C., Kolter, R. (2005) *Nat Cell Biol* **7**, 933–934.

Wagner, V. E., Bushnell, D., Passador, L., Brooks, A. I., Iglewski, B. H. (2003) *J Bacteriol* **185**, 2080–2095.

Waters, C. M., Bassler, B. L. (2005) *Annu Rev Cell Dev Biol* **21**, 319–346.

Whiteley, M., Lee, K. M., Greenberg, E. P. (1999) *Proc Natl Acad Sci USA* **96**, 13904–13909.

Wu, H., Song, Z., Givskov, M., Doring, G., Worlitzsch, D., Mathee, K., Rygaard, J., Hoiby, N. (2001) *Microbiology* **147**, 1105–1113.

Wu, H., Song, Z., Hentzer, M., Andersen, J. B., Molin, S., Givskov, M., Hoiby, N. (2004) *J Antimicrob Chemother* **53**, 1054–1061.

Yang, F., Wang, L. H., Wang, J., Dong, Y. H., Hu, J. Y., Zhang, L. H. (2005) *FEBS Lett* **579**, 3713–3717.

Index

Note: The letters 't' And 'f' following locators denote tables and figures respectively.

S.T. Reddy (ed.), *Paraoxonases in Inflammation, Infection, and Toxicology*, Advances in Experimental Medicine and Biology 660, DOI 10.1007/978-1-60761-350-3, © Humana Press, a part of Springer Science+Business Media, LLC 2010